T0312883

JANA SCHAICH BORG,
WALTER SINNOTT-ARMSTRONG,
AND VINCENT CONITZER

Moral AI
And How We
Get There

A PELICAN BOOK

PELICAN
an imprint of
PENGUIN BOOKS

PELICAN BOOKS

UK | USA | Canada | Ireland | Australia
India | New Zealand | South Africa

Penguin Books is part of the Penguin Random
House group of companies whose addresses can
be found at global.penguinrandomhouse.com.

First published 2024
001

Text copyright © Jana Schaich Borg, Walter
Sinnott-Armstrong and Vincent Conitzer, 2024

The moral right of the author has been asserted

Book design by Matthew Young
Set in 11/16.13 pt FreightText Pro
Typeset by Jouve (UK), Milton Keynes
Printed and bound in Great Britain by
Clays Ltd, Elcograf S.p.A.

The authorized representative in the EEA is
Penguin Random House Ireland, Morrison
Chambers, 32 Nassau Street, Dublin DO2 YH68

A CIP catalogue record for this book is available
from the British Library

ISBN: 978-0-241-45474-9

Penguin Random House is committed to a
sustainable future for our business, our readers
and our planet. This book is made from Forest
Stewardship Council® certified paper.

www.greenpenguin.co.uk

To our children,
whose world will be shaped by AI:

Dalia, Skye, and Aiden

Miranda and Nasa

Naomi, Ben, and Aaron

Contents

Preface

The three of us come from different fields: Jana is primarily a neuroscientist and data scientist, Walter is primarily a philosopher and ethicist, and Vince is primarily a computer scientist and game theorist. It is not easy for researchers from such diverse fields to meet or, when they do meet, to understand each other. Luckily, our group was not deterred by these difficulties. We chose to work together because we all share deep concerns about ethical issues raised by artificial intelligence (AI) and want not only to understand these problems but also to provide actionable suggestions about how to nourish AI innovation while minimizing AI's moral harms. No single field by itself could overcome such complex challenges.

We also share optimism about AI, although we agree that AI development must be pursued thoughtfully and responsibly. Whereas many others depict AI as all good or all bad, we recognize that AI has great potential for both good and harm. This book is our attempt to explain the positive and negative effects of AI in a balanced way so as to enable AI's benefits at the same time as reducing its dangers.

Our collaboration started as a series of informal conversations, which quickly morphed into a research group that at various times included Duke University undergraduates (especially Rachel Freedman in computer science and Max Kramer in philosophy), postbacs from various institutions (especially Kenzie Doyle), graduate students in business (Daniela Goya-Tochetto) and computer science (Vijay Keswani, Yuan Deng and Duncan McElfresh), philosophy postdocs (Gus Skorburg, Lok Chan, and Kyle Boerstler), and additional faculty members (John Dickerson and Hoda Heidari in computer science). The group continues to meet regularly and frequently recruits new members from different fields.

We are deeply grateful for all the help we received from these collaborators and from many friends and students – too many to list – for useful comments and discussions on our papers and talks. For comments on the manuscript, we thank Rachel Freedman, Alice Wong, Hoda Heidari, Ben Eva, and students in a graduate seminar on Philosophy of AI. We also owe a tremendous debt to Casiana Ionita and Penguin Press for their encouragement and patience when the COVID-19 pandemic slowed us down. And, of course, we thank our family members who compassionately put up with us when we dedicated time to this book, despite pressures of the pandemic and three new family additions.

Some of these chapters are based on, and reprint modified parts of, previously published articles and chapters. In particular, Chapter 6 is based in part on 'How AI can aid bioethics' by Walter Sinnott-Armstrong and Gus Skorburg, *Journal of Practical Ethics* 9(1) (2021). Chapter 7 includes some ideas published in 'Four investment areas for ethical AI: transdisciplinary opportunities to close the publication-to-practice gap' by Jana Schaich Borg, *Big Data and Society* 8(2) (2021) and 'The AI field needs translational ethical AI research' by Jana Schaich Borg, *AI Magazine* 43(3) (2022). We thank the publishers and co-authors of those articles for permission to revise and reuse that material here.

Finally, we are very appreciative of financial support for our research from Duke Bass Connections, Duke Collaboratories (awarded by Duke's Associate Provost for Interdisciplinary Studies), the Future of Life Institute, the Templeton World Charity Foundation, the National Science Foundation, and OpenAI. Without all of you, this book would have been impossible. Of course, none of these funders or supporters is responsible in any way for the content of this book. That responsibility is ours alone.

Introduction:
What's the problem?

'The robots are coming! The robots are coming!' This alarm is spreading quickly, partly because of popular movies like the *Matrix* series, which spreads fear that AI will soon control everything. Even those who are not terrified of losing their freedom to AI masters still might worry that AI will rob us of privacy and make it too dangerous for us to live our personal lives as we please, that it will make it impossible to know whether an email or picture is fake, that it will manipulate us into voting for malevolent political candidates or buying consumer goods that we do not really need, or that it will harm our lives in countless other ways. To avoid any or all of these dangers, such pessimists argue that we must stop – or at least slow down – AI before it is too late.

Others are much more optimistic. They look forward to the day when AI will perform tasks that humans are bad at or want to avoid. Instead of fearing self-driving cars, these types of AI optimists fear drunk human drivers and are inspired by predictions of how many deadly accidents autonomous vehicles could prevent. Instead of fearing killer drones, they seek to end the mistakes and intentional atrocities that human soldiers commit out of intense emotion, ignorance, or sleep deprivation. They believe that AI will help us make rapid progress in science and medicine which will improve human lives. Some even look forward to the idea of humans *merging* with advanced AI to become a vastly more capable species. To gain these benefits, such optimists oppose any attempt to slow down progress in AI development.

Which is it? Should we be pessimistic or optimistic about AI? Is the glass half empty or half full? Our answer is: *both*. There is both bad news that gives us reason to be worried about some uses of AI and good news that gives us reason to advocate for other uses. Indeed, there is sometimes reason for fear *and* hope about the very same AI technique or application. So AI deserves both pessimism and optimism. For example, consider the following areas where AI is being used.

TRANSPORTATION

Some good news: Steve Mahan, who is blind, uses his autonomous car to get around Santa Clara, California, sometimes alone by himself.[1]

Some bad news: Joshua Brown was killed in May 2016, when his Tesla Model S in Autopilot mode crashed into a white truck that it could not distinguish from the bright sky.[2]

MILITARY

Some good news: An AI-augmented bomb-disposal robot can access tight spots and defuse or detonate deadly devices without requiring humans to risk their lives.[3]

Some bad news: In Lohatlha, South Africa, an Oerlikon GDF-005 anti-aircraft gun went out of control when directed by AI and sprayed hundreds of cannon shells, killing nine and wounding 14.[4]

POLITICS

Some good news: AI can be used to draw congressional districts in North Carolina that are fair to both political parties and various interest groups.[5]

Some bad news: Cambridge Analytica used AI trained on the personal data of up to 87 million Facebook users without their knowledge in an attempt to influence votes in the 2016 United States presidential election.[6]

LAW

Some good news: Thought River created an AI-based program that is supposed to read legal contracts, answer key questions about them, and suggest what to do next, all for a much smaller fee than human lawyers charge.[7]

Some bad news: Eric Loomis was sentenced to six years in prison partly because an AI-driven risk assessment identified him as a 'high risk to the community'. It is not clear how the AI makes its prediction, because its algorithm is proprietary, but critics argue that it is biased against some racial and gender groups, including groups Eric Loomis was a part of.[8]

MEDICINE

Some good news: In 2015, IBM's AI, Watson, diagnosed a female patient with a rare leukaemia that human doctors couldn't pinpoint initially. The doctors claimed, 'We would have arrived at the same conclusion by manually going through the data, but Watson's speed is crucial in the treatment of leukemia, which progresses rapidly and can cause complications.'[9]

Some bad news: Bias in data used to train an AI to predict

the need for 'high-risk care' led it to dramatically prioritize White patients over Black patients even when they had the exact same level of illness.[10]

INVESTMENT

Some good news: AI-driven financial advisers and investment tools, such as those provided by Betterment, Wealthfront, and Wealthsimple, claim to be at least as accurate as human advisers. Due to their low account minimums and transaction costs, such robo-advisers allow lower-income investors to get financial advice that was previously available only to the wealthy and privileged.[11]

Some bad news: On 6 May 2010 at 2:45 pm, the Dow Jones Industrials Average dropped 998.5 points within a few minutes due to AI-driven stock trading systems engaging in high-speed automated selling spirals that crashed the market.[12]

MARKETING

Some good news: Have You Heard? (HYH) used AI to increase the number of homeless youths who get tested for HIV by 25 per cent through identifying young people who were likely to be more influential in their social networks.[13]

Some bad news: In 2012, a Target store sent coupons for baby items to a 16-year-old customer because her buying patterns had led their AI system to predict that she was pregnant. Her father complained to the store before she told him that she really was expecting.[14]

ART

Some good news: Pandora and Spotify (among others) use AI analyses of musical taste to recommend new songs and artists that listeners enjoy and might never have discovered without such services.[15]

Some bad news: AI has been used to produce a great deal of impressive art, which has sold for as much as $432,500.[16] However, the AI is able to do that because it was trained on images created by human artists that were used without the artists' permission, and without compensation or credit. Now that art-generating AIs have been created, it is technically difficult – and in some cases impossible – to force them to 'forget' the images of human artists who do not want their art included, even if the artists' work is copyrighted.

MEDIA

Some good news: The *Los Angeles Times* uses an algorithm called Quakebot to warn its readers about California earthquakes more quickly and accurately than traditional reporting.[17]

Some bad news: In 2023, Muharrem Ince, a candidate for president of Turkey, withdrew from the race because a sex video of him was posted on Facebook which he claimed was a 'deepfake', an AI-generated picture of something that never happened. In the same presidential race, a fake video was circulated of a terrorist organization supporting Kemal Kılıçdaroğlu, another candidate. 'When internet users turned to Google to search for Kılıçdaroğlu on that day, the false news was among the top suggestions made by the algorithm.'[18]

SURVEILLANCE

Some good news: AI has been used successfully to locate poachers in India and Africa, and even to predict poaching before it happens so that park rangers can intervene. Game-theoretic AI can also optimize ranger patrol routes in order to catch poachers or intercept the snares they set out.[19]

Some bad news: The Chinese government is 'using a vast, secret system of advanced facial recognition technology to track and control the Uighurs, a largely Muslim minority'.[20] Other countries are also reported to use facial recognition AI for similar targeted surveillance.[21]

ENVIRONMENT

Some good news: Blue River Technology uses AI-driven camera technology to 'see' the ground, distinguish crops from weeds, and then deliver targeted herbicides that kill only the weeds. The technology is claimed to help avoid herbicide resistance and save cotton growers money on herbicides.[22]

Some bad news: Tremendous computational resources are needed to train many high-performing AI models. It has been said that 'training a single AI model can emit as much carbon as five cars in their lifetimes'.[23]

These examples are just the tip of the iceberg. You can probably think of many more yourself. Still, this sample illustrates the types of multifaceted issues that society and AI creators need to grapple with and that we will discuss in this book.

To think through AI's potential advantages and dangers

to society, we need to be clear on the nature and limits of AI. When does written code become AI as opposed to just a computer program? Is artificial intelligence really intelligence? Can AI be conscious? Can AI be inventive and creative or can it only mindlessly follow the instructions of its creators? Can AI perform new and different tasks or can it only perform the same old tasks that humans have performed before? Which kinds of tasks can AI do – and which kinds can AI not do? Can AI become far more intelligent than human beings? We will tackle these questions in the next chapter.

Once we understand what AI is and is not, we must consider how AI interacts with basic moral values, including:

Safety – as in cases of autonomous cars and weapons, deepfakes, social media, robot surgeons, and future AI systems that could cause humans to lose control over our world or parts of it.

Equality – as in cases of criminal risk assessments based on race, wealth, and gender, bias in healthcare provision, robo-advisers increasing access to financial services, and rising economic inequality from job losses due to AI.

Privacy – as in cases of Cambridge Analytica, Target's coupons based on pregnancy predictions, and Chinese surveillance of Uighurs.

Freedom – as in the cases of aiding blind people like Steve Mahan and impeding movement or religious practices through targeted surveillance.

Transparency – as in the cases of Eric Loomis and of AI programs that grade job performance, résumés, creditworthiness for loans, and student papers.

Deception – as in cases of deepfakes and AI-generated fake news used to interfere in elections.

Some AI uses implicate more than one value, and this list of values is not meant to be exhaustive. But even this initial list shows how wide-ranging and serious the ethical issues associated with AI can be.

The dangers of AI should not be *under*estimated, but they also should not be *over*estimated. Typically, AI can be built and used safely and ethically as long as the types of moral issues listed above are addressed thoughtfully. To keep the AI baby without its bathwater, we will try to illuminate the moral issues at stake and show why we all need to pay more attention to AI ethics.

CHAPTER 1
What is AI?

Many people's ideas of AI are shaped by blockbuster movies. If you have watched *Star Trek*, you might think AI is a benevolent, human-looking supercomputer that has trouble processing and understanding emotions. Most people have also heard that AI is behind our phones' ability to transcribe what we say or to answer our questions, sometimes with a comical lack of success. Are all of these things truly AI? Does something have to look like a human and be impressively intelligent to be considered AI, or can AI be just anything that does something cool or helpful for us, even if it doesn't always do it well?

If you're hoping that the world's best AI researchers have settled on a clear definition of what AI is, you will be disappointed. Definitions tend to be abstract and vary greatly. When John McCarthy, an American computer scientist and cognitive scientist, first coined the term 'Artificial Intelligence' in 1955, he defined it as 'the science and engineering of making intelligent machines', and said, 'Intelligence is the computational part of the ability to achieve goals in the world.'[1] Since then, some definitions of AI suggest that AI must perform tasks that normally require human-level intelligence or even use the same cognitive computations humans do while achieving those tasks, while other definitions make no claims about how AI relates to human intelligence, but specify that AI must somehow learn over time. Some people in the field use an 'I know it when I see it' approach to define AI, because they want the AI label to connote some minimum level of algorithmic or task complexity. A common result of this definition is a phenomenon called the *AI effect*: once we know how to write a computer program that can perform a task we once thought required intelligence, such as navigating mazes, we no longer think the computer program's solution to the task is *really* intelligence. Instead, it's 'just' an algorithm. As a result, some experts in the field no longer care what it means to be 'intelligent' in a broad sense, because they believe that the future of AI lies in performing individual tasks exceptionally well using whatever method works best, even if the most effective method doesn't seem particularly smart or insightful.

For the purposes of this book, we have agreed to use the following very broad definition of AI, which is a slightly modified version of the one provided by the US National Artificial Intelligence Initiative Act of 2020:[2]

> A machine-based system that can, for a given set of human-defined objectives, make predictions, recommendations or decisions influencing real or virtual environments *with sufficient reliability* [we added the clause in italics].

The fact that it is machine-based is the sense in which it is artificial. The fact that it reliably pursues objectives is supposed to make it intelligent (though we will need to discuss intelligence much more below).

This definition is quite permissive, but we wanted it to

have breadth because many (although not all) of the ethical issues surrounding AI that we want to discuss exist no matter whether we consider mundane AI used in familiar industry applications, cutting-edge AI achieving superhuman performance on specific tasks, or human-level general-purpose AI as it may be developed in the future. Relatedly, most of the strategies we propose to address these issues do not require intermediate steps of determining exactly how intelligent an AI system is. That doesn't mean there isn't valid debate about what a system – biological or artificial – needs to do to qualify as intelligent. It just means that we won't attempt to settle that debate here. It is important to acknowledge that AI systems have different goals with varying degrees of ambition, though.

'Narrow', 'general', 'strong' – pick your challenge

Most kinds of AI that companies use and that we encounter in everyday life are examples of 'narrow' AI, which is AI that is designed to perform a specific task (or a small set of tasks) using whatever computational strategy its human designer thought would work best. This is in contrast to 'general' AI, which could perform a great variety of tasks, possibly including ones for which it was not designed. It is a topic of debate to what extent we already have general AI systems. Some systems are at least getting close. For example, advanced chatbots – like ChatGPT from OpenAI, Bing Chat from Microsoft, Bard from Google, and Claude from Anthropic – can answer many types of questions and perform many types of language-based tasks, including ones that their creators never thought to ask them. These chatbots are based on what are called *large language models* (LLMs) that have been trained on enormous amounts of text from which they can learn many kinds of things. Still, these chatbots are incapable of, say, driving a car, so in this sense they are narrow. At the same time, increasingly, the field is discovering techniques that work across many problems, so that very similar techniques may be at the core of multiple systems that do very different things. Even if those systems individually are narrow AI systems, the underlying techniques are general-purpose and can plausibly contribute to general AI.

Sometimes people also use the term 'artificial general intelligence', or 'AGI'. This term tends to have connotations of human-level AI (or beyond), meaning that it is at least as good as us at basically *any* cognitive task. Some people dislike the term 'AGI', preferring instead to talk about 'human-level' AI, or 'superintelligence' that far exceeds us. Normally, though, there is little difference in how these terms are used, especially by people who think that if human-level AI is reached, then superintelligence can't be far off. It is clear that we do not yet have such AI systems today, but some people think we could see AGI emerge in our lifetimes.

A different distinction that is sometimes used is between 'weak' and 'strong' AI. Different people take this distinction to mean different things. Some interpret it as equivalent to the narrow-versus-general distinction, typically with weak AI corresponding to narrow AI and strong AI corresponding to AGI. Others take weak AI to comprise computational tools meant to *mimic* specific aspects of human intelligence, and strong AI to be (general) AI that functions using cognitive states *identical to*, or at least closely resembling, those of humans. This allows for a distinction between, on the one hand, researchers who pursue general AI but don't have a primary aim of making their AI systems function like humans, and, on the other hand, researchers pursuing general AI who are more motivated to make AI think like humans.

Among general and strong AI enthusiasts, there is disagreement about whether general or strong AI should be, or needs to be, conscious like most humans are. This is further complicated by the fact that, even for humans, our understanding of consciousness (including what that even is) remains very limited. Partly for this reason, most AI researchers sidestep the question of consciousness. We will do the same here. That said, the question of whether AI systems might be conscious has already made its way into the news. Blake Lemoine, a Google engineer, argued that Google's LaMDA (Language Model for Dialogue Applications) was sentient.[3] (Philosophers often take sentience and consciousness to be closely related.) Most AI researchers are very sceptical of this claim, but the claim will be hard to definitively disprove as long as we understand consciousness so poorly. Since AI systems keep getting better, we can be sure we haven't heard the last of the AI consciousness question.

Before continuing, we want to warn you that the term 'weak' AI is pretty misleading. Despite its name's connotation, weak AI can be extremely impressive and can dramatically outperform humans at a given task. For example, Deep Blue, the system that beat then-world champion Garry Kasparov at chess in 1997, was a form of weak AI.[4] It worked by evaluating 100–300 million possible moves per second, and choosing the best option of the moves evaluated according to rules defined and fine-tuned by AI engineers and chess masters. For the most part, it didn't learn or update its rules in an automatic way, and it certainly didn't play chess the way humans do. Nonetheless, by being able to perform certain kinds of computations much faster than humans can, it was able to beat the best human chess player in the world. It took decades for computer scientists to figure out how to achieve that milestone, and Kasparov reported that he sometimes saw such 'unusual creativity' in Deep Blue's moves that he assumed a human was intervening on Deep Blue's behalf.

Despite the celebrated success, Deep Blue was retired to the Computer History Museum almost immediately after beating Kasparov. Why? Because its creators knew Deep Blue couldn't do anything other than play chess. This is the definition of a system being narrow, and even those who think 'weak' means something different from 'narrow' will agree that this is a weak AI system because it doesn't play chess using the same kinds of multipurpose cognitive mechanisms humans do. But let's not forget that this old, weak AI can beat the pants off us humans at the specific task it was designed to do. Similarly, it's important not to fall into the trap of thinking that weak AI can't do very complex things that, when pursued by humans, draw deeply on humans' innate intelligence. All these forms of AI can be impressive in the right – or sometimes wrong – contexts. Don't let the field's currently preferred names fool you.

A good word for trivia night: GOFAI

Another term worth knowing is 'GOFAI', which stands for 'Good Old-Fashioned AI'. Sometimes when people say 'symbolic AI' they mean roughly the same thing. It works by doing computations about human-cultivated symbolic representations of its 'world' according to rules or inference procedures predetermined by its programmers. That's a mouthful, but we can understand the definition through some examples. Deep Blue's GOFAI was given a way to represent the states that a chessboard can be in, the moves that can be taken from those states, which new states those moves result in, the states in which the game is over, and who (if anyone) has won when the game is over. It was also given a way to estimate how good a particular state of the board was likely to be for it. Based on this, Deep Blue could then perform the following computation: *search through sequences of moves to find moves that are likely to get the board to a better state (even though the opponent is working against this)*. As another example, GOFAI floor-cleaning robots may be given a way to represent facts about their world ('I am in the southwest corner of the room', 'When there is dark stuff in a corner of a room, that means the corner is dirty', 'There is dark stuff in the northwest corner of the room'), actions that the robot can take ('Move forward one metre'), a specification of how taking that action changes the facts about the world, and goals that the robot should pursue ('The entire room should be clean'). Based on this, it can then perform the following computation: *reason logically to figure out how to clean the room*.

GOFAI systems are used in many industries and can be very powerful. For instance, many military robots and weapons use GOFAI, and many planning, logistics, and scheduling systems use a mixture of GOFAI and other techniques. Some AI researchers continue to believe that the only way to make artificial systems that are truly intelligent is to use GOFAI in the process.

The problem with GOFAI is that if you don't know the exact symbols, rules, and logic to program into a system to adequately describe the situations it will encounter, your system will have a high probability of not working well. That got GOFAI researchers into trouble for a couple of decades, because it turns out to be very difficult to specify *everything* you, as a human being, do to perform tasks in the real world. Consider a task that GOFAI is actually quite good for: scheduling a set of meetings so that everyone can attend the meetings they're supposed to be in. Most GOFAI scheduling systems are designed to take into account people's scheduling constraints and room availability, because that's usually the most important information a system needs to schedule meetings successfully. But not all the time. Imagine that an employee requests a meeting with their manager's boss to discuss their manager's inappropriate behaviour. The last person the employee will want to see on their way to or from that meeting is their manager, because the manager might wonder why the employee is talking to their boss, suspect the employee is complaining about them, and maybe even retaliate. Nonetheless, the system schedules the meeting for the time slot right before their manager meets with their boss because that's what makes most efficient use of everybody's calendars. As a result, the employee and offending manager end up having a very awkward encounter in the hallway in front of the boss's office. A human scheduler might have anticipated this encounter and scheduled the meetings differently, but without explicitly giving the GOFAI system representations of human social dynamics, notification of what the meetings are about, and guidelines for how the topics should impact the schedule, the GOFAI scheduling software would be blind to the problem. This type of unforeseen error tends to mean that GOFAI systems need to be carefully limited in their scope and overseen by humans who understand their limitations. Some people use the GOFAI label with a hint of derision because of these constraints.

Now let's consider a task GOFAI approaches aren't a good fit for. In the American game show *Game of Games*, contestants have to state which celebrity is in a random picture. This is not exactly the kind of problem that you would expect on an intelligence test, but it is something humans do easily and something automatic facial recognition programs are designed to do, too. One of the challenges of this task is that celebrities appear in different clothes, in different settings, from different angles, and under different lighting conditions. They may even change their hair colour or get plastic surgery to modify their appearance. So how would you create a GOFAI to identify celebrities? What rules would you create? 'If the pixels in the top left quadrant are mostly yellow and there are some blue pixels in the middle and . . . [lots of other conditions] . . . then it is a picture of Ellen De-Generes.' Maybe someday we will learn enough about the rules the human brain uses to recognize people to make this work.

In the meantime, the GOFAI approach seems hopeless for this problem.

Teaching machines to learn

What can we do instead? We need an approach for completing tasks that doesn't require us to know explicitly how the tasks should be accomplished. We can draw inspiration from human cognition. Humans learn from experience. Everything we encounter as we go through our lives somehow gets integrated into what we know, even though we don't always know how. Can machines do the same thing? This is the idea behind a type of AI called 'machine learning'.

Machine learning gives AI a goal, and then lets it figure out by itself how to achieve that goal through experience. 'Experience', in this case, means encountering or processing a lot of data relevant to the goal. Machine learning algorithms use that data to build some type of model or set of representations that allows them to make predictions or decisions about new data. The model-building process is usually referred to as model 'training' or 'learning'.

The machine learning field has developed a couple of main approaches. The first is often used when we know things in advance about the data that we are using to train our machine learning models. For example, if a group of friends were to look at the pictures used for *Game of Games*, they could probably figure out between them which celebrity is in each picture. If we compensate them with enough pizza and brownies, we could probably convince the group of friends to go through each picture and label it with the celebrity it contains. This labelled data could then be given to an AI system to learn what various celebrities look like, so that it could label pictures of these celebrities by itself in the future. This approach to machine learning is called *supervised* learning. It's a teaching strategy similar to teaching children using flashcards or through stating the names of different animals as you point to them on a farm. Supervised learning is arguably the most common type of machine learning, and it is the most straightforward non-symbolic approach for making predictions. When you use supervised learning, you train the AI with a labelled data set that allows the AI to create a model of the relationships between the features of the data set and the associated labels, and it uses that model to make predictions about new, unlabelled data.

In other cases, it's too difficult to get labels for data or we don't even know what labels we should try to use. Then we need to use *unsupervised* machine learning methods. Unsupervised learning doesn't need labels and instead focuses on finding patterns. It looks for commonalities between data and makes classifications or predictions based on whether those commonalities are present. This is somewhat like the unconscious learning we do all day long as we observe the environment in the background of whatever tasks we are focusing on. Some common unsupervised machine learning techniques include k-means clustering and principal component analysis. These techniques might produce, for example, the category groups news aggregator apps use to organize articles about similar stories from different news sources. Some unsupervised learning paradigms are designed to be 'generative' – that is, they try to figure out the patterns that describe the data they learn from well enough that they can create new data that is hard to tell apart from their training data. These are the types of methods that are typically used to power chatbots that talk with you, create fake pictures that look real, or generate AI-created art.

Two potential limitations of both supervised and unsupervised approaches to machine learning – as well as variants such as semi-supervised learning, which uses both labelled and unlabelled data – are that, first, the AI's success is usually very dependent on whether the data on which it was trained is representative of the situations in which it will actually have to perform. Second, most standard approaches tend to make predictions in a static way. That is, the data used for training represent static snapshots of the world and predictions based on these data are largely treated as being independent from one another. In real life, though, sometimes we have to learn or figure out what to do without having much data about what are good actions to take or what actions have worked well in the past, and without being able to cleanly separate things into snapshots. Consider walking. When a baby starts to try to walk, they have no prior personal experience of walking to guide their movements. They only know that the people around them seem to be able to move quickly on two legs and they want to be able to do that, too. So babies try and fall, and try and fall, and try and fall . . . until they figure out how to move their feet in a way that allows them to get from one part of the room to the other.

This 'try and fail' ('try and fall', in the case of walking) type of learning experience is the paradigm inspiration for another machine learning approach called *reinforcement* learning. The primary goal of reinforcement learning is to maximize some type of reward by learning from the successes and failures of many attempts over time. To frame learning how to walk as a reinforcement problem, we might specify that reaching the other part of a room gives a reward of 10 while falling down receives a reward of –3. A standard reinforcement learning AI is not given explicit instructions about how the world works, so it must figure out by itself through trial and error how to maximize its reward, even though its interactions with the surrounding environment can change which actions will lead to reward over time. (For example, going to a certain blueberry bush may be a good way to get blueberries at first, but after you eat all the blueberries from that bush, it's no longer a good bush to choose.) As a consequence, reinforcement learning AIs start out being terrible at a task but take advantage of their failures to get better and better over time.

Reinforcement learning does not require labelled data, but it does require the AI to interact with an environment, which it typically starts out knowing little about. Therefore, reinforcement learning is best suited to settings where the AI is perpetually doing something it can learn from, like driving an autonomous car or playing a video game. Between those two examples, the case of playing a video game has the obvious advantage that no real damage is done while the AI is still learning. Likewise, one strategy AI engineers use to help reinforcement learning AIs gain experience is to have an AI run huge numbers of simulations, sometimes competing against itself in those simulations. This approach is credited with helping AlphaGo, the AI created by Google's DeepMind, to beat one of the world's best Go players, Lee Sedol, in 2016.

What about deep learning and neural networks?

If you have read anything about AI in the news recently, you probably have heard the terms 'deep learning' and 'neural networks'. How do these fit into the AI types and terms we have discussed so far?

Every approach to machine learning must summarize what it has learned into some kind of representation that allows it to respond to new inputs from the environment with the appropriate output. One popular type of representation, whose fundamental ideas have been around since the 1940s, is called a *neural network*. Neural networks represent information through input, hidden, and output layers of artificial neurons or 'nodes' that are connected to each other. Nodes are computational units of different kinds, whose design was inspired by neurons in the human brain. Like the brain, the connections between nodes have varying strengths that change over time. Nodes' activations are determined by the presence and strength of 'activity' in connected nodes in earlier layers, and the strength of the connections to those nodes. Similar to human brain parts that have neural layers or areas with unique functions, artificial neural networks are designed so that individual nodes each process information with their own thresholds and modifying factors, resulting in different parts of the artificial neural network doing different things.

Artificial neural networks with lots of hidden layers are called 'deep' neural networks, and learning achieved using these networks is called 'deep learning'. Deep learning started to outperform many other techniques in the early 2010s, first in visual recognition tasks and later in many others, including natural language processing, or the automated analysis of text written in human languages. This somewhat sudden success after so many years of trailing other AI methods was in part due to the development of new architectures in deep learning, including the 'transformer' model, which involves attention layers in the network that can pay attention to 'far away' input, such as a word much earlier in a piece of text. Deep learning's take-off was also due to advances in computer hardware that allowed the networks to be scaled to unprecedented, and ever growing, sizes. Scaling these models requires a combination of drastically increasing (in the order of billions) model parameters, training data, and computational power to get the model trained up efficiently, so computer hardware can cause a real bottleneck in progress. Intriguingly, these models seem to acquire more sophisticated abilities as they are scaled further, as illustrated by OpenAI's sequence of GPT models that have performed better and better as parameters, training data, and computational power are added. Believe it or not, scientists do not yet understand why scaling neural networks can impact their performance so drastically. As a result, there is somewhat of a fervour to build the largest neural networks possible to see what unexpected capabilities they might be able to learn, and you hear about these attempts a lot in the news.

Sometimes people get confused and think neural networks and deep learning are types of AI, distinct from supervised, unsupervised, or reinforcement learning. This isn't quite right. Rather, neural networks provide a way to represent data that supervised, unsupervised, or reinforcement learning can employ.

Which AIs are intelligent?

We have touched on only a few of the most visible approaches in the AI field. There are many others, plus we left out many details and variants of the ones we did discuss, and novel approaches are constantly under development. Why do we need so many different types of AI?

Well, for better or for worse, we don't yet know how to tell a computer program 'Go be intelligent!' and have the outcome be anything useful. At least for the time being, the methods we have access to require us to articulate specific problems or tasks that the AI is meant to solve in a mathematical fashion. Figuring out how to do this is sometimes the hardest part of developing AI. Another big challenge is finding relevant data that will allow AI to solve the problems we ask it to solve. The data that are available or feasible to collect are often not the data we would choose in an ideal world. Further, while some entities have the financial resources to access, store, and process massive data, others do not. That then brings us to our answer: at our current level of technological development, we still need a variety of different types of AI to address the different types of problems we would like AIs to solve and to accommodate the different types of data we have access to.

Each AI approach has its own advantages and disadvantages. For example, unsupervised learning puts few constraints on what kind of data you need to have, which can make an AI developer's life easier. At the same time, it tends to require tremendous amounts of whatever data you do have, and it gives you less control over the outcome, which may not be what you want. In the end, engineers choose the type of AI they think is best suited to the type of problem they want to solve and the data they have.

Often, for the purpose of real-world deployment of AI systems, the most successful results come from combining multiple different AI kinds, tools, or models into what we will refer to as 'AI systems'. In fact, many AI researchers believe combining AI approaches is the only feasible way to advance AI functionality to the level we want, and they have designed approaches that do this formally. For example, while AlphaGo relied on deep learning techniques, achieving its level of play also still required searching through ways that the game might play out (which is more like a GOFAI technique), so deep learning alone was not enough to beat top human players. Similarly, self-driving cars use different modules for different tasks, rather than being controlled by one single AI model or tool. One module may focus on perceiving the surroundings, another on making decisions about how to drive, etc., and each of these modules may use a different type of AI.[5]

Despite the practical uses of these types of heterogeneous AI systems, some of the world's best researchers hold strong beliefs that GOFAI will have no significant role in building general AI, and that general AI will be accomplished primarily through advances in deep learning, with the role of other AI systems limited to tools that the general AI sometimes chooses to make use of. Others believe that there's no way to create a truly intelligent system without symbolic AI or

GOFAI. The debates about which AI approach is best will likely rage on.

No matter which approach or technique is used, though, there will be reputable experts who question whether the system it contributes to should really be considered artificial *intelligence*, as opposed to just a computing system that performs a complex task well. To understand why, we need to explain what today's best AI systems are not so good at.

What today's AIs tend to lack

AI is already outperforming humans in a variety of narrow domains – and not only games like chess and Go. So, what advantages do humans still have? What are the major gaps between today's AI and human intelligence? The answer to that question is changing rapidly, but here are a few of the gaps that have historically been difficult to close. Some argue that large language models used by the most advanced chatbots have bridged some of these gaps, but as we will discuss later, this is debatable in ways that can be important. We'll start with some high-level weaknesses and then make our way into more technical issues.

THEY LACK COMMON SENSE

AI is often bizarrely poor at integrating the type of information that allows it to reason about basic, everyday phenomena. Consider GPT-3, an older version of the large language models that ChatGPT is based on. When GPT-3 was given a prompt like 'What is the most important piece of advice on creativity?' it responded with impressively coherent paragraphs of text that made it seem like GPT-3 might really understand the question and its answer. Here's an example:

> I think creativity is a talent and it's something you can learn. I think the most important advice is to nurture your curiosity. Get to know yourself. Don't censor yourself. Don't hide your ideas. Express yourself and surprise yourself. Check out more of the work, here.[6]

Upon further inspection, though, it becomes clear that while GPT-3 picked up on many patterns in human language, its actual understanding of what it was saying was still very limited. For example, when GPT-3 was asked what to do about a situation in which someone needed to move a table through a door, but the table was too wide for the doorway, GPT-3 advised that they should cut the door in half and remove the top half.[7] If you're not really paying attention, superficially it sounds like a plan that might work. But thinking about it for a second, there's no reason to think that cutting the top half of the door will help you move the table – not to mention that you probably don't want to damage your house this way! GPT-3's answer lacks basic common sense. This problem plagues many AIs, even though more recent large language models seem to have fewer common-sense failures than GPT-3 did.

THEY ARE ONE-TRICK PONIES

Most of today's AI systems do not have any broad representation or understanding of how the world works beyond the specific problem or task they were designed to solve. This is why pretty much every example of successful AI is narrow AI. The AI systems have no broader framework to draw on to guide them in how to address tasks they weren't explicitly designed and trained to complete.

In fairness, some recent systems do seem to be going in a more general-purpose direction. Today's large language models can produce text about anything and be used for many diverse tasks that depend on text prediction (like language translation, chatbots, or document summaries). In another sense, though, they are still examples of narrow AI because the only thing they can do is predict text. They can't perform a different type of task, like drive a car.

Some cutting-edge models can learn and accomplish many tasks at once, like DeepMind's Gato, which performs 604 diverse tasks, ranging from captioning images to playing Atari games, to stacking blocks with a real robot arm.[8] Nonetheless, Gato still can't do things that are dramatically different from those 604 tasks it was trained for (and, at least for now, it's not particularly good at those tasks anyway).[9] Most AIs do not, in any case, have Gato's task fluidity. Human beings, on the other hand, know about a lot of different things, and we can use our knowledge flexibly to address new scenarios or tasks we have never encountered before. We do this all day long, usually with ease.

THEY CAN'T THINK 'OUTSIDE THE BOX'

Pablo Picasso famously stated that calculating machines are 'useless because they can only give you answers'.[10] Human creativity requires more than just giving answers or just giving logical outputs from a set of inputs. Some AI systems exhibit at least some form of creativity. For example, there are AIs that create music or poetry that many customers can't distinguish from humans' music or poetry,[11] and AlphaGo played brilliant moves in the game of Go that no human would ever have played. Still, these AI systems have historically been creative *only within the box that they are given*, and their creativity is restricted by the data they have been trained on. For example, AlphaGo never suggested a change to the rules of the game of Go, such as introducing a new type of stone that would make the game more interesting. Large language models don't avoid the restrictions they were programmed with by drawing images instead of printing text.

For readers with children of any age, in contrast, you know that even very young humans naturally think outside the confines of a task given to them, in order to get what they want. Toddlers, kids, and teenagers constantly come up with ingenious ways parents never anticipate, to avoid doing what they have been asked to do without strictly breaking their parents' rules – even though they are well aware what the parents intended when they set the rules. It's as if they are trying to change the *de facto* constraints of a game to their advantage without changing the formal rules they were given. AIs sometimes also complete tasks we give them in ways we don't intend (a phenomenon called 'specification gaming'), but unlike with humans, the AI system doesn't have any better idea about what its designers wanted than a literal interpretation of the rules or goals given to it. In this way, it can be argued, artificial creativity is still more constrained than human creativity.

THEY STRUGGLE WITH HIERARCHICAL PLANNING

Imagine you have decided to visit Amsterdam. You start making a travel plan. First, you consider which dates work for you. Then you look for flights and a room. Maybe you schedule some specific activities. Closer to the date of departure, you flesh out your plan in a bit more detail. You plan what to bring and how to make it to the airport in time. On the day of departure, you plan even more detailed steps. When you get to the airport, you figure out which line to stand in to check your luggage. Then you choose the best way to get to your departure gate. This continues all the way until the end of your trip, and usually you succeed in making it home.

This way of proceeding is entirely natural to us, and it's hard to even imagine how else you might plan your trip. Nonetheless, there is an alternative. In principle, we *could* approach the entire planning problem from the perspective of moving individual parts of our body. 'Which body part should I move first and how? Which part should I move next and how? Which part should I move next and how?' Following the plan might entail steps like using your finger to press down on a key on the keyboard (as part of booking a flight), lifting your left foot off the ground (as part of walking in the security line), then your right foot, and so on.

Although this approach might work *in principle*, it is immediately obvious that it would be a very poor strategy for planning a trip to Amsterdam. First, it is inefficient. Why spend the time outlining every move your body can make when you could instead just say something more general, like 'I'll go through the security line'? Second, the number of potential sequences of muscle movements you could consider is mind-bogglingly large, so even the fastest computer on the planet could not possibly explore them all. Third, it seems unlikely that the strategy would work, because it's almost impossible to anticipate all the issues that could impact which movements you should make. What if there is traffic on the road, lines at security, delayed flights, or people and objects you have to manoeuvre around at the airport? You can't realistically plan detailed movements for every contingency ahead of time, and so you would have to redo your plan as surprises emerge. In contrast, a normal human plan provides a high-level roadmap, allowing you to say, 'OK, so it turns out the train isn't running, but the point of getting on the train was just to get to the airport and I can do that by driving a car instead.'

Believe it or not, the inefficient planning strategy we just described is similar to the way many AI systems work. They often have only a very limited set of options to approach problems with (like 'determine the next motor movement'). It is hard to create AI planning systems that think abstractly about which flights to take without worrying about all the details of every movement needed to pack clothes or how to find your passport in the bottom of your bag when you need to show it. Unlike traditional planning AIs, a large language model like ChatGPT can in fact produce text containing a high-level plan. However, ChatGPT cannot unpack those instructions into the detailed actions you need to take to get into your car and out of the garage. The instructions it gives you are useful for your general planning, but you still need to be able to translate the higher-level travel plan into individual movements to actually make the trip to Europe happen. Humans go back and forth between these different ways of planning for the trip automatically and with ease. AIs can usually only plan at one level of abstraction, and it is challenging to help them think 'hierarchically' or choose different ways of abstracting or representing parts of a problem to make the problem as a whole more manageable.

When AIs have enough processing power and the world in which they function is sufficiently constrained, like the world of virtual chess, the make-my-singular-approach-to-solving-problems-work-through-brute-force strategy might succeed. When AIs have to function in the real world, though, where there are almost infinite possibilities of how the environment will manifest at any given time, even the most highly resourced AIs may not have the speed and processing power for the brute-force approach to work. The result is that it is surprisingly challenging to create AIs that can make their way to Europe on their own (or that do things like clean and tidy up after a party). AI planning is an exciting area of research which is making progress, but for now most AIs still struggle (or completely fail) to plan in complex environments.

THEY LACK EMOTIONAL AND SOCIAL INSIGHT

Humans know what it is like to be a human being, and, at least on our better days, we are good at empathizing with other human beings. We can often glean a lot of information from what somebody is saying, not because they're directly stating that information, but through inference and perception. For example, we consider how the person generally feels about other things, we hear the person's intonation, we see the person's body language, we pay attention to the fact that the person chooses to use particular words, we can imagine how we would feel in that situation, etc. We are generally not even aware of all the reasoning we do to gauge how someone feels about something. That has made it challenging to replicate such reasoning in algorithms. It's also not clear what kind of data is best to feed into AIs to help them learn about social information and, in particular, social interactions. So, today's AIs are not as socially or emotionally intelligent as we are.

AI is making significant inroads here. There are many AIs trained to identify which emotion a person is expressing in a picture, in a voice recording, or in their written text. There have also been successes in training AIs to anticipate what specific other actors will do and for what reasons, which some argue is similar to human theory of mind, or the ability to understand that other beings have their own unique goals, plans, and mental states.[12] Still, the performance of these systems can be very inconsistent (even when companies selling these systems claim otherwise), and they aren't yet able to pick up on or navigate the same level of social and cultural complexity that many humans are. That's why, despite excitement and investment in creating AI-driven virtual companions, psychiatrists, or teachers, few people report feeling the same emotional connection or ease with those AIs as they do with their human counterparts.

THEY USUALLY CAN'T LEARN FROM A FEW EXAMPLES

When AI systems outperform humans, they are typically trained on far more data than a human will ever get. For example, AlphaGo's successor, AlphaZero, played 21 million training games. It required only 34 hours to complete its training – programming tricks allowed much of this training to be done in parallel, rather than playing the games sequentially like humans would need to do. For humans to accomplish the same amount of training, they would have to live to be 100 years old, play Go every day from birth, and somehow play more than 50 games per day. Humans just don't have that kind of time. That's why humans have evolved to be able to learn many things after only having been exposed to them once. AIs still struggle with this kind of 'one-shot' or 'few-shot' learning. Here again, there is progress. For example, large language models can often learn new tasks with only a few examples. Even so, the models only achieve this 'few-shot learning' after being 'pre-trained' on immense amounts of data, way more data than humans ever require.

Well, you might say, it's too bad that humans can't practically do the same amount of training as AI can, but why force AI to abide by human constraints? If AI can train far more than we can in a reasonable amount of time, then that is an advantage of AI, not a weakness. Treating it like a weakness is kind of like a sore loser saying, 'Usain Bolt runs much faster than me, but I'm not impressed – he wouldn't be able to do that if he had *my* legs!'

There is a practical reason that AI's need for immense amounts of training stunts AI's potential, though. When we're talking about learning to play chess, Go, or even a game such as *Starcraft*, it is relatively easy and cheap to train for millions of games. All the training takes place in simulation, and if the AI plays miserably in some of this training, no harm done.

In contrast, it is far harder and more costly to let self-driving cars take millions of *real* cross-country trips to learn to drive. Further, we can't just let the self-driving cars explore freely and try all possible actions to find out what happens ('What if I just speed up when I see a red light?'). If we allowed them to do that, society would rebel against self-driving cars very quickly. Even more often, we simply don't have enough data to train the AI in the task we want, and have no idea how to go about collecting more data. Data availability and cleanliness are a huge (and expensive) bottleneck in today's AI ecosystem that will persist until AIs can learn more efficiently through fewer examples. Humans don't have this problem, at least not to the same degree.

THEY CAN'T INTERACT WELL WITH THE PHYSICAL WORLD

Even when AIs can generate good abstract, high-level plans for doing something in the world, actually going out there and interacting with the physical world is a different challenge. Robotics is the field dedicated to making machines that can replace human (or animal) physical movements. To achieve this, robots need ways to sense what's happening in their environment, perform physical actions, and adjudicate between the potential physical actions they can take at any given moment. You may have seen videos of robots doing amazing things, but those same robots often fail in spectacular ways, and many physical engineering challenges continue to make building machines that successfully navigate the physical world very hard. Most robots still can't grasp small objects like spoons or do tasks with other entities collaboratively, like carry a table with another person, despite countless efforts and tremendous investment.

Another formidable task is playing football – the kind that Americans call soccer. If you ever want an amusing way to procrastinate, watch some YouTube videos of the Robo-Cup competitions. The stated eventual goal of the RoboCup competition is to have a team of fully autonomous humanoid robots beat the winner of the most recent World Cup, complying with all of FIFA's rules. Despite this lofty goal, when teams in the RoboCup humanoid league (the league whose robot competitors most closely resemble human beings) play today, you might question whether the robots are even playing soccer. The robots very, very cautiously and slowly shuffle across the field, occasionally gently tap the ball in the right direction, and frequently fall over. Sometimes all the robots get confused about what to do, and the robots stay still for tens of seconds. It's fun and even cute to watch, but Lionel Messi is not worried about AI-driven robots taking his job anytime soon. That is not to say that the RoboCup initiative will not succeed by 2050. Watching YouTube RoboCup videos of competitions across the years, you can see dramatic progress that far exceeds progress in the human game. But the human game is definitely still far ahead.

It is worth mentioning here that some AI researchers and neuroscientists believe AI will never be able to reach human-level intelligence without having a physical body in the world, mainly because the human brain learns a tremendous amount through physical perception of sensations within our own bodies. The jury is still out on whether that's the case, but if so, the challenges of robotics will end up being challenges for human-level general AI as well.

THEY ARE VULNERABLE TO CONTEXT CHANGES

One overall consequence of the specific strengths and limitations described above is that today's AI algorithms can perform well in structured, repetitive environments. The more predictable the environment, the more potential AI has to perform better than humans. That's why AI has done so well providing diagnoses based on medical images or identifying signs on the road (a problem that self-driving cars need to solve). The more unstructured or variable the environment that a task needs to be performed in, though, the worse AI tends to do. Without some kind of fundamental knowledge about the world, common sense, or ability to reason in a principled abstract way, learning based solely on associations in lots of data is brittle. As soon as small things change in the data that are being collected or in the contexts that predictions need to be made in – perhaps because images the AI uses are suddenly taken in new weather conditions or with a different device – performance usually drops significantly, often in very unexpected ways.

Sometimes performance drops because of another general limitation of AI that we've already alluded to: it doesn't generalize or adapt to new tasks well. You might think, that's OK, I don't adapt immediately to new tasks either! The fact that I know how to drive a car doesn't mean I know how to drive a motorcycle or motor boat. But when humans know how to drive a car, we more quickly and more easily learn how to drive a boat because at least some of the skills needed to drive cars and motor boats are the same. Both require steering, controlling one's speed, and keeping an appropriate distance from those around you, for instance. The human learning system naturally transfers these relevant skills from one context to the other.

AI systems, on the other hand, often struggle to do this. As counterintuitive as it may seem, many machine learning systems need to be trained up all over again from scratch to perform even a slightly different task than the one they were originally designed to address. Avoiding retraining from scratch on every new task is an important topic in machine learning, referred to as *transfer learning*. While there has been significant progress on that, humans still do it far better.

Another reason performance can drop when AI operates in unpredictable environments can be a little more concerning. When AIs learn to perform tasks purely through patterns in data, as tends to be the case with machine learning, rather than through fundamental knowledge of the world, sometimes they end up performing the task well through associations that have little fundamental relationship to the task. As a consequence, the AIs suddenly perform horribly when new contexts cause the learned spurious associations to change.

In a famous example, one team trained up a deep neural network to be able to discern the difference between a picture of a husky and a picture of a wolf, a task some humans might find challenging.[13] At first the neural network did great. Then suddenly it didn't. Why? Because most of the husky pictures the neural network was trained on had snow in the background (since huskies are often favoured dogs in colder environments). Few wolf pictures the neural network was trained on had snow in the background. As a result, the easiest and most effective correlations in the data that differentiated between the huskies and wolves ended up being those related to which colours were in the background of the picture. The neural network basically ending up being a white background detector! As soon as the system was given a picture of a husky on a non-snowy background, it failed. This kind of thing happens very often in machine learning. Humans' performance of similar tasks tends to be much more robust in different environments and settings.

Well, AI's tendencies to make predictions through spurious correlations and conceptually irrelevant relationships wouldn't be too problematic if there were a straightforward way to track how and why the AIs were making their decisions, right? After all, even though humans make plenty of mistakes, we can often prevent ourselves from making those mistakes in the future if we or a teacher can help us understand why our reasoning was faulty. Similarly, if we knew exactly how AI systems made their decisions, we could quickly catch what was going wrong in a different environment or even anticipate the problem in advance. That brings us to the next limitation of today's AIs.

THEY CAN BE DIFFICULT TO INTERPRET

Unfortunately, many of the best-performing AIs, especially deep learning AIs, function as 'black boxes', at least in a sense. It takes a lot of work to get insight into what led the AI to choose a certain output given a set of inputs. That makes it arduous, and sometimes impossible, to figure out how to prevent the AI from making similar mistakes again or to predict which other kinds of mistakes the AI is likely to make in the future.

Of course, it's important to acknowledge that humans can't always explain their decisions either. If someone shows you a picture of your mother, you'll recognize it instantly. If that person asks you to explain how you know it's her, you might struggle to answer. Perhaps you will try a description like, 'Well, she has blue eyes, and her eyes are about this far apart, and. . .' but your description will still match many people that you, personally, would never mistake for your mother. Unless there is something particularly conspicuous about your mother, like an unusual scar or a unique hat that you know she is wearing, you probably would not be able to give your friend adequate verbal instructions to help them pick out your mother in a large crowd, even though you could pick her out instantly.

So, is it too much to ask an AI system to be explainable or interpretable? If humans sometimes can't explain their choices, why should we expect AI systems to do so? Well, there are a few reasons. First, although humans *sometimes* can't explain their choices, they just as often *are* able to explain their choices, and to adjust them quickly in response to feedback about their reasoning. Second, it's usually possible for us to understand why humans make their mistakes, even if we wouldn't make them ourselves. AI's mistakes, on the other hand, are often perplexing and can be some real headscratchers. When IBM's Watson took part in the TV game show *Jeopardy*, it responded with 'What is Toronto?' (a Canadian city) in response to a clue in the 'US Cities' category. This would be an extremely odd response from a human being with a reasonable grasp of North American geography. It was very surprising coming from Watson, given how much relevant information it had access to and was trained on. To this day, nobody is entirely sure why Watson gave that response. Maybe it didn't realize the response had to be a US city? Maybe Watson knew of the rural community called Toronto in Illinois? It's a mystery. The worry is that if we can't understand why an AI system makes such a mistake, then it's difficult to prevent similar mistakes in the future or to feel confident that the AI system won't make other unanticipated, possibly extremely harmful, mistakes if it is deployed.

This worry is even more significant when you realize that sometimes this inability to predict mistakes can be exploited, intentionally getting an AI to make a specific mistake. For example, researchers managed to fool a face recognition system simply by one of them putting on specially designed eyeglass frames. The face recognition system then consistently labelled their researcher as Milla Jovovich, even though the researcher looked nothing like her to human eyes.[14] While that example is amusing, there are more nefarious uses of such techniques; for example, imagine someone tricking an AI system into falsely diagnosing a politician as having a serious medical problem, which in turn harms her career prospects and leads to unnecessary treatment.

It is getting harder to know what AIs can't do

From all this, it may appear that we are still quite far away from human-level AI. If each of AI's limitations described

33

in this chapter requires a separate breakthrough, the field still has a lot of breakthroughs to make! That's a reasonable assessment. At the same time, many of these shortcomings may be related to each other, so it's possible that a single breakthrough will take care of more than one limitation. For example, maybe we could figure out how to give AIs a broad understanding of the world, and, in doing so, provide AIs a way to learn from a few examples and be able to transfer information across separate, but related, tasks. At this point we just don't know how many fundamental advances will be needed to make human-level AI a reality or to allow narrow AI to overcome its current challenges. What we do know is that we don't know how to create human-level AI *yet*, but AI progress has been tremendous and unpredictable, and anything is still possible.

The past few years have also put the field in a new and interesting predicament. Even if most (or all) AIs have the limitations described above, technology has advanced so much that some AIs do a really good job of, at the very least, *making it seem* like they don't have those limitations. Large language models provide the most obvious instances of this phenomenon. The basic idea behind large language models is that they use neural networks to predict which words would be acceptable in the next spot in a sentence, given the context. Those predictions can then be used to allow the AI to do almost any task where language needs to be used, like the tasks chatbot assistants are given. The models have become so advanced that they often generate responses that, if taken at face value, suggest that AIs have succeeded in overcoming many of the limitations we discussed earlier. For example, chatbots might now be able to give you a basic plan for travelling from one place to another ('Go to Times Square and take the subway to Queens'), and, when appropriately prompted, they may be able to build out this plan a little further ('If you don't already have one, you can buy a MetroCard at the station'). This sounds a lot like hierarchical planning, perhaps even creative hierarchical planning. Chatbots are sometimes also able to accurately describe what the characters in a story are feeling or are likely to do. This sounds a lot like emotional or social intelligence. But do these chatbots *really* have the characteristics of creativity, hierarchical planning, social intelligence, and so on?

It is common to be sceptical. The models are just good prediction machines, by this view. There's nothing inherent in their construction that allows them to 'understand' the critical concepts referenced in the sentences they generate. Rather, they just have enough statistical knowledge about correlations between words in the text they were trained on to spit out things that sound good. As some like to say, the language models know the *form* of language, but not its *meaning*. But not everybody agrees.[15] Further, in one survey, a slight majority of computer scientists working on human language said they thought large language models *of the future* could someday 'understand natural language in some nontrivial sense' if they are given enough data and computational resources.[16] So how are we to think about this? And what does this mean about the differences between human and AI capabilities?

Humans can represent information in a different way than large language models were designed for. Humans can use causal models of concepts – categories, situations, and events that mean something in the real world beyond their statistical properties in language. In fact, we seem to have an inherent drive for this type of understanding of things around us. We will call this the 'mental model' way of functioning. Large language models were designed to represent information in a different way. They are fed a whole bunch of text and learn correlations between what words are shown together in different contexts within that text. Their computational power allows them to exploit the statistical correlations they learn. We'll call this the 'statistical correlation' approach to functioning.

When people say that a human 'understands' something, they usually imply that the human has accurate conceptual models of that thing, even if humans also learn many things through the statistical correlation approach. In contrast, even those who want to claim that large language models 'understand' things agree that they understand things through the statistical correlation approach. It is much less clear whether large language models can use or understand concepts. It is certainly *possible* that the large language models have somehow created hidden causal models within their webs of learned correlations which function like concepts, but we don't yet have any idea how to test for the existence of them. Also important, even if conceptual models are embedded somewhere within neural networks, we have no idea how to access them. In general, techniques for interpreting what is going on inside neural networks are still very limited. So we are left to speculate about what large language models actually know, while grappling with intuitions about 'understanding' that seemed in the past to make sense when applied to humans but that fall short when we try to use them on these AI models.

How different are the mental model and statistical correlation approaches to knowledge? And when do the differences matter? At this point we really don't know, which means we also don't have adequate ways to think about what large language models understand or the ways they understand. This also means it is getting increasingly difficult to describe the gaps between human and AI abilities, and what AI systems still struggle to do.

We have done our best to describe the gaps as they stand now, but we know this account may be completely outdated in ten years, or even five years. On the other hand, it wouldn't be that surprising if these gaps persist either. Only time will tell.

Who makes today's AI?

Before diving into AI's ethical implications, there is one more aspect of AI it is important to understand: how it is made. Science fiction depictions of dangerous AI often assume AI is a robot created in the basement of an evil genius. In reality, unless that evil genius had access to a tremendous amount of data and was extraordinarily educated and independently wealthy, they would need help from many evil geniuses and probably many rich evil investors as well. Most things that use AI require contributions and resources from a lot of different people, even when they employ very narrow AI for a specific purpose.

Things that use AI are created through at least three distinct stages. First, AI *algorithms* are developed to describe which functions and logics will be applied to data to create the AI's intended behaviour. AI algorithms are typically developed in some kind of research environment (academic, corporate, or government), often by people with doctoral degrees in highly mathematical fields.

Next, at least in applications that use machine learning, AI *models* are 'trained' by passing carefully chosen data through AI algorithms. AI models represent what was actually learned by an algorithm, and provide outlines of what factors the AI will take into consideration and how. AI model training requires more data wrangling and software engineering than AI algorithm development, and is completed in a wide variety of settings by data scientists, engineers, computer scientists, analysts, or even enthusiasts from the general community who may or may not have advanced mathematical expertise.

Most of us do not interact directly with AI models or algorithms, though. Instead, we interact with AI *products*. AI products are the composite experiences, interfaces, or devices that allow us to interact with trained AI models or algorithms for a specific purpose. They can be websites, phone apps, recommendations, robots, drones, or any of the other examples we will discuss in this book. AI products require business, design, and operational expertise to develop, and usually considerable financial investment as well. Thus, they are most frequently developed by teams within businesses and organizations.

AI product teams typically involve user experience (UX) researchers, user interface designers, many types of engineers, business analysts, and managers. User experience researchers figure out which features products need in order to attract users, user interface designers figure out what those features should look like, engineers create the products according to other team members' specifications, and business analysts and strategists make sure the product is supported by a profitable (or at least sustainable) financial model. A product manager (or product 'owner') sets the strategy for what the product will ultimately do and what kinds of users it will target. They also steer the team by facilitating communication between the researchers, designers, engineers, and the organizations sponsoring the creation of the product, and by prioritizing workflow in a way that balances the needs of all the stakeholders. Other contributors, such as lawyers, compliance officers, data scientists, or AI specialists, may also help create a product and sometimes do so through second parties or consulting services. Notably, the people who create AI algorithms themselves are often *not* part of an AI product team or even part of a product's sponsoring organization at all. Some of the most advanced AI technology companies may have AI researchers or experts with cutting-edge backgrounds collaborating with some product teams, but this scenario is the exception rather than the rule. Many companies even use AI through AI-as-a-service (AIaaS) providers, meaning they pay another company for access to AI tools like language translation models or face recognition models. As we will discuss, these dynamics within the AI creation process have many implications for AI's impact on society, so keep them in mind as you read on.

Now we know what AI is and at least have an idea of how it's made. It's exciting stuff! It's also tough stuff. Of course, the technical and practical issues of creating AI are just a small subset of the challenges we need to overcome for AI to make our lives better. We also need to tackle AI's ethical challenges. Let's dig in to find out what those ethical challenges are.

CHAPTER 2

Can AI be safe?

*'The development of full artificial intelligence could spell
the end of the human race. . .'*
— Stephen Hawking on the BBC

*'The pace of progress in artificial intelligence (I'm not referring
to narrow AI) is incredibly fast . . . The risk of something
seriously dangerous happening is in the five-year time frame.
10 years at most.'*
— Elon Musk in a comment on Edge.org

*'I don't want to really scare you, but it was alarming how
many people I talked to who are highly placed people in AI
who have retreats that are sort of "bug out" houses, to which
they could flee if it all hits the fan.'*
— James Barrat in the *Washington Post*

Thoughtful AI advocates, despite being enthusiastic about
all AI has to offer, justifiably fear that it often won't be safe –
where 'safety' is understood as freedom from unacceptable
risk or harm. Of course, many other useful technologies pose
safety concerns, too. Nuclear power plants provide power
without emitting large amounts of greenhouse gases, but
they can release cancer-causing radioactive contamination

when damaged. Medical devices or drugs often treat one ailment, but they also introduce the risk of life-threatening infections and side effects. Missiles can prevent or stop wars, but they also kill innocent civilians. Should AI be used to benefit society only with significant restrictive safety guards? Or, alternatively, is AI safe enough that we should make it as available as possible for all people to use without limiting constraints? We explore these questions in this chapter.

Will life with AI become a scary sci-fi fantasy?

We begin with the most familiar and extreme safety concern: AI that takes over the human world. In the *Terminator* movie series, a military AI system called Skynet launches nuclear missiles to prevent humans from trying to unplug it. A human resistance force then spends many full-length movies trying to defeat Skynet. (After a lot of heroics, they ultimately succeed, thankfully.) Should we worry about AI that is so intelligent that it dominates and outmanoeuvres humans? The people quoted at the top of this chapter think we should, and they aren't the only ones. Many serious AI researchers do, too.

Here's the scenario AI researchers worry about the most: we give a superintelligent AI – meaning an AI that is smarter than humans across the board – instructions to do something, but we fail to anticipate the consequences of the AI following our instructions. Philosopher Nick Bostrom illustrates this with the following well-known example.[1] Suppose we are in the business of making paperclips. One day, we obtain a superintelligent AI system and decide to let it take care of the paperclip production process. We instruct the AI system to 'maximize the number of paperclips produced'. Sounds both sensible and harmless, right? However, since our instructions to the AI don't include anything about when to stop making paperclips, the AI continues to try to make paperclips even if we no longer need them. Moreover, it tries to make paperclips in the most effective way possible. It embarks on turning the Earth into one incredibly efficient paperclip factory. It even launches a space colonization programme to turn other planets into paperclip factories too. We try to stop it, but the AI anticipates our actions and outmanoeuvres us at every step to make sure we don't get in the way of its paperclip-making goals. Pretty quickly, the AI decides that humans are making it harder for it to do its assigned job and, as a result, launches nuclear bombs to eradicate us. The AI was never programmed to be evil and never had an explicit goal to harm us, but due to our inability to anticipate the consequences of our original instructions, humans and other life are nonetheless eradicated from Earth. Some AI researchers refer to this as 'the King Midas problem'. King Midas was initially ecstatic when his wish that everything he touched would turn to gold was granted, but he discovered his folly when all the food he tried to eat turned to gold,[2] and, in a later version of the legend, he tragically turned his daughter into gold while embracing her. Similarly with AI, if you try to be specific about what you want, you are likely to leave out some important details, and that can result in many outcomes that you think are terrible.[3]

Assume the AI could become intelligent enough and physically able to lauch nuclear bombs to eradicate us. This does not mean we need to have given it direct access to the nuclear weapons; a sufficiently intelligent system could probably gain access on its own by a combination of manipulating humans and hacking other computer systems. It would be hard to stop a superintelligent system just by trying to restrict its physical abilities or access to other systems, at least as long as humans interact with it in some way. (And what would be the point of building an AI system we don't even interact with?) But we should still be able to stop the AI before it launches nuclear bombs, right? After all, we humans are the ones creating the AI! That's a reasonable hope, and it is possible that we could intervene in time, but doing so requires humans to be able to detect or anticipate that the AI's capabilities are getting to a worrisome point.

The problem is that it's not clear we can be relied on to do that. First, if humans don't yet realize how powerful an AI's intelligence is, they won't have any reason to take adequate steps to contain the AI. Second, increases in AI capabilities may happen too quickly for intervention. In theory, one lone researcher could come up with a new idea in a single day which immediately results in a big enough boost in intelligence for an existing AI system to become dangerous and unstoppable. A twist on this idea is that we create an AI system to design other AI systems for us, and it quickly creates ever more advanced AI, so an artificial intelligence explosion takes off that leaves us in the dust and is now out of our control. And third, even if there were some clear signals that AI systems' abilities are getting to a dangerous level, we might still not stop improving them. A company making such AI systems might agree with the concern in principle, but see other companies pushing ahead, and decide it has to do the same to stay competitive.

You could argue that we are seeing this already: today, there are lots of concerns about large language models, even within the companies producing them, and yet companies continue to race to have the best LLM system. Sometimes this is even motivated by the thinking that *our* company will best be able to control these systems and mitigate the danger! The same dynamic can play out between countries; many people talk about an AI race between the US and China, for example. This line of thinking is easy to fall into.

Are these terrifying scenarios just sci-fi fantasy? Many AI researchers believe that we are not *that* close to human-level AI, for the reasons described in Chapter 1, but super-intelligent AI is still at least theoretically possible. Actually, that theoretical possibility is what inspired many to join the AI field in the first place – the idea that intelligence of any kind can be built into computers. Still, there is little agreement about how long it would take to generate human-level or superhuman-level AI, or how likely it is to come about. Some have argued that it is pointless to worry about AI superintelligence at this point, like AI researcher Andrew Ng, who in 2015 compared worrying about 'evil' AI to worrying about overpopulation on Mars.[4]

Others take the current pace of technology as reason to be genuinely concerned; confronted with the abilities of large language models, for example, many researchers are changing their minds about how far these techniques can go. Even if superintelligent AI doesn't exist today, it seems fathomable that superintelligent AI could exist within some amount of time on the order of decades. The AI field has already progressed much further than most people expected, and there have already been unexpected, dramatic advances. In particular, the vast majority of the AI community did not see the deep learning revolution coming at all, which is notable, since many of the highest-performing AI systems currently utilize deep learning. Even during the past decade when the potential of deep learning became clear, many AI researchers underestimated over and over again how quickly these techniques would gain new capabilities. Anticipating more unpredictable breakthroughs in the future, it seems plausible that today's children will encounter human-level or superintelligent AI in their lifetimes. We worry about what the state of the climate will be in 2100, as presumably we should. So shouldn't we also worry about future superintelligent AI?

In our view, it is worth thinking about the dangers of superintelligent AI and at least starting to plan a containment strategy for them. For example, one approach is creating AI that does not need to be given an explicit goal, such as maximizing paperclips, but instead cautiously learns from us what we really want (similar to an approach we will propose in Chapter 6). We think that society should invest in technology that learns and implements human values, along with thoughtful discussion about which policy or organizational practices are needed to ensure that superintelligent AI, if it is ever developed, does not harm us. Some of these practices will be discussed in Chapter 7.

But these problems are very challenging. One issue is that, for now, we don't have any superintelligent AI systems to test our techniques on. And the worry is that once we do, it's already too late to do anything. Meanwhile, even today's, non-superintelligent AI systems are extremely difficult to control. For example, OpenAI as a company thinks a lot about how to get its systems to do what we really want them to – often called the 'alignment problem'. But when its systems, such as ChatGPT, are made available to the world, it quickly becomes clear that OpenAI's precautions still fail in lots of ways – not least because users try to get them to fail. For example, OpenAI is in an arms race with users trying to 'jailbreak' ChatGPT to produce text, code, instructions, etc., that it is not supposed to produce.

We could write a whole separate book on the safety concerns of superintelligent AI. In the meantime, it is essential to consider the safety issues of the AIs that already exist. Even today's AIs could pose an existential threat to us if, for example, they get to control nuclear warheads or discover something about physics, chemistry, or biology that we were unaware of but which could pose a threat to us. (AI is already involved in making scientific discoveries: for example, Deep-Mind's AlphaFold is dramatically better at predicting protein structure than anyone or anything that came before it.) Yet these safety issues are not fully appreciated or managed, especially when advanced AI technology is intentionally made available for anyone to use. It is these contemporary issues that we will turn to now. Not only are these issues clearly important in their own right, but they can also teach us some things about controlling AI systems more generally, and they have the benefit of being tangible and concrete.

Safety concerns surrounding today's AI

Concerns about AI superintelligence grow out of fears of what AI can potentially do when it becomes far more capable than it is today. The safety concerns related to today's AI systems, in contrast, usually grow out of recognition of what AI still does poorly and the challenges of teaching humans with diverse backgrounds how to use AI properly in applications where we don't yet know how AI will perform. Some of the present-day issues related to those safety challenges arise from the many ways in which contemporary AI is currently limited, as well as misunderstood and misused, by humans. Here is a quick list of these problems:

IT'S INTELLIGENT, BUT THAT DOESN'T MEAN IT WON'T MAKE MISTAKES ('AI MISTAKES')

It's easy to think that 'artificial' intelligence means perfect intelligence, in part because we're not used to computers making mistakes when we use them for simple calculations, like those on a calculator. However, most AI systems in use today are not particularly intelligent. Almost all of them will sometimes make incorrect predictions or take actions that do not help them achieve their objective, and some of them make those types of mistakes often. It can be surprising to people to see a system like ChatGPT appear so intelligent on one problem, and then fail on something that to us seems so obvious. The problem is that when we see how it does on the first problem, we start thinking about which *humans* can solve that problem. Those humans are quite intelligent, so we are tempted to think that ChatGPT must be just as intelligent. But that's not how it works. When AI systems make mistakes in contexts where they are assumed to be highly accurate, many kinds of harm can ensue. It's tempting to think that this type of safety issue is easy to address simply by making all AI systems overridable and overseeable by human experts. However, that leads us to the next safety issue.

HUMANS MAY TRUST AI TOO MUCH, EVEN WHEN IT MAKES MISTAKES ('TOO MUCH TRUST')

People often make decisions as if AI systems are more accurate than their performance justifies. For example, the Dutch tax authority blindly trusted an AI to determine whether families applying for childcare allowance were committing fraud, even though the AI was known to make incorrect predictions, leading to a large number of needy families being unfairly fined, especially in certain demographic groups.[5] Many studies show that humans can't be relied on to override AI mistakes in these situations. Moreover, when humans do successfully monitor AI's mistakes, it can lead to the next safety problem.

AI'S POTENTIAL FOR MISTAKES CAN DISTRACT HUMAN DECISION-MAKERS ('HUMAN DISTRACTION')

Monitoring AI systems can take a lot of cognitive resources and energy, especially when it is known that an AI model is only moderately accurate. As a consequence, AI tools can actually cause human decision-makers to perform *worse* in contexts that require extreme concentration and fast choices, such as when AIs give false alarms during surgery or in air traffic control systems.

AI'S USERS AND MONITORS MIGHT MAKE MISTAKES
('HUMAN MISTAKES')

We typically want humans to be able to monitor or override AI so they can take corrective actions when they detect mistakes. However, if humans are not sufficiently knowledgeable or trained in how AI will behave when humans take such actions, their attempts to correct the situation can be undermined, and sometimes even be completely counterproductive. You may have experienced something similar with anti-lock brakes (even though standard anti-lock brakes don't use AI). It is completely normal for anti-lock brakes to shudder when you brake suddenly; but if you are not aware that the shuddering is normal, your natural inclination might be to assume that something is wrong with the brakes and release your foot from the brake pedal, which is the exact opposite of what you originally intended to do.

AI'S SUCCESSES MAY DETERIORATE HUMAN EXPERTS'
SKILLS ('SKILL DETERIORATION' OR 'DESKILLING')

Many human skills deteriorate if they aren't practised on a consistent basis.[6] If AI frequently replaces that practice, the humans who are supposed to quality-check AI lose the skills they need to evaluate AI effectively. For example, our spatial memory during navigation becomes impaired proportionately to the amount of time we use AI-driven Global Positioning System (GPS) tools.[7] Thus, in some sense, the more we use AI-driven GPS tools, the less qualified we become to evaluate them. Indeed, we often become completely unaware that our relevant skills have been affected.

UNDETECTED AI BIAS CAN AFFECT THE QUALITY OF LIFE OF SOME GROUPS MORE THAN OTHERS ('HARM BIAS')

We will discuss the ways in which AI is often biased in Chapter 4. Bias becomes a safety issue when it affects which people or groups get harmed, including who lives or dies.

AI ENABLES BAD INTENTIONS ('EMPOWERING EVIL')

AI can be used for tremendous good. In the wrong hands, it can also do tremendous harm. The more available AI is to those with nefarious intentions, the more effectively those nefarious intentions will be carried out. Scammers can, for example, use AI to predict your account passwords and steal your money.[8] Large language models can enable a flood of hard-to-detect bot accounts that take over discussions on social media.[9]

AI SUPPORTS NOVEL WAYS TO CAUSE HARM ON MASSIVE SCALES ('NEW TYPES OF HARM')

AI allows people with nefarious intentions to wreak havoc in brand-new ways that were never available before. A few examples are AI-individualized phishing attacks, AI-generated 'deepfakes' that can convince people to hand over money or information, AI-driven behavioural manipulation that can convince people to vote in a certain way, and AI adversarial attacks (more on this later).

AI REACTS TO OTHER AI TOO QUICKLY FOR US TO CONTROL ('AI–AI INTERACTIONS')

Sometimes the best way to address an AI safety issue is to create a different AI that will address it. In contexts like high-frequency trading and cybersecurity, different parties each deploy their own AI systems, and all of them then interact in the same domain. This leads to AI-versus-AI ecosystems where AIs are making decisions on our behalf at incomprehensible speeds. These interactions can lead to surprising – and potentially devastating – outcomes faster than we can possibly stop them. An early example was the 2010 trillion-dollar 'flash crash' precipitated by high-frequency traders.[10]

AI FACILITATES PSYCHOLOGICAL DISTANCE ('AI-MEDIATED DEHUMANIZATION')

The less we view the object of harm as 'human', the less empathy we feel towards them and the less we disapprove of causing harm to them. The more 'psychologically distant' we feel from others, the easier it is for us to dehumanize them. AI can contribute to psychological distance in many ways, but one of the most fundamental is predisposing us to treat people as numbers instead of humans. To illustrate, when people were told a COVID-19 hospitalization and death-toll dashboard was generated by an AI instead of human engineers, they were less likely to support preventive safety measures to protect others, and their level of support correlated directly with the extent to which they viewed the dashboard's data as data about real people versus cold numbers or statistics.[11]

Some case studies

These safety issues might seem abstract, so let's look at some examples of how they manifest in concrete situations. We will start by thinking through known 'safety-critical' applications where it is well accepted that malfunctioning and mistakes can lead to death, serious physical and mental injury, or severe damage to property or environment. Then we will move to domains that are not typically believed to be 'safety-critical', but where AI still has the potential to cause significant levels of injury and harm.

AI IN TRANSPORTATION

Approximately 1.3 million people die each year from automobile crashes, and traffic deaths disproportionately affect developing countries.[12] The United Nations set a goal to reduce traffic injuries by 50 per cent by 2030, and it is widely acknowledged that AI technology will be critical for reaching this goal.[13] As a result, investment in AI-based cars, trains, planes, and even personal jetpacks has exploded in recent years. Lots of terrific progress has been made. Of course, as helpful as transportation AI can be, it will inevitably make and cause mistakes that need to be reduced to acceptable levels.

Aeroplane crashes are less common than car crashes, but they harm more people at once. Two Boeing 737 MAX plane crashes in 2018–2019 occurred because of AI transportation mistakes. Both crashes were caused by an automated system called the Manoeuvring Characteristics Augmentation System, or MCAS, gone rogue. The purpose of MCAS was to automatically push down the nose of the plane when sensors at the front of the plane determined the plane was flying at too steep an angle, possibly putting the plane at risk of stalling in midair. In both crashes, the sensors malfunctioned and made the system think it should push the nose of the plane down when it shouldn't have ('AI mistakes').

The pilots did not know how to turn MCAS off and were not able to adjust the other settings in the aeroplane to overcome MCAS's action ('human mistakes'). As a result, the planes were plunged into the ground while the pilots were desperately trying to point the planes back into the air, killing 346 people.

Neither of the Boeing 737 MAX crashes would have happened if MCAS had been more easily overridable or if the pilots had been sufficiently trained in how and when to override it. In addition, some people might argue that MCAS doesn't meet definitions of AI that require the system to learn over time.[14] Perhaps if MCAS had learned over time like many other types of AI systems, at least the second crash could have been avoided. Still, the Boeing 737 MAX crashes are a reminder of how many lives can be lost without such training and learning.

But now let's imagine that such training and learning were accomplished, and MCAS sensors were improved so that the MCAS was more trustworthy. Moreover, imagine that the system was extended to take over more functions from the pilots. In this scenario, 'deskilling' becomes a real concern. The longer humans rely on autonomous vehicles, the less reliable their own driving or piloting skills will become, making them increasingly less reliable in emergency situations. This is already a documented issue in the aviation industry. A pilot's adroitness at bringing a Boeing jet with a crippled engine in for a landing in rough weather correlates closely with how much time the pilot has *recently* spent flying planes manually, as opposed to with the help of automated systems.[15] According to one researcher, 'Flying skills decay quite rapidly toward the fringes of "tolerable" performance without relatively frequent practice."[16] As a result, many pilots have seemingly forgotten how to take over a flight manually if the automated systems fail, or they do not know how to adapt to an automated system that is malfunctioning or turned off – the purported causes of accidents such as the Air France Flight 447 in disaster 2009[17] and the Asiana Airlines crash in San Francisco in 2013.[18]

What about cars? It's easy to imagine how lack of practice would have a similar impact on typical automobile drivers as well, making them unable to correct an automatic vehicle's mistakes in emergencies. Many of us have already started to feel this kind of impact by becoming increasingly uncomfortable parallel parking without rear-view cameras or by becoming overly reliant on automated indicators that tell us if a car is next to us when we change lanes.

Another problem is that self-driving cars can make humans less vigilant. In Arizona in 2018, a pedestrian called Elaine Herzberg was killed by a self-driving car operated by Rafaela Vasquez. The self-driving car's sensors detected the pedestrian 5.6 seconds before impact but never treated her as something to avoid and continued on its path. Unfortunately, Vasquez, who had been operating – or at least riding in – the vehicle without incident for over 40 minutes had stopped paying close enough attention. She was distracted and looking away from the road, so she didn't intervene until too late to prevent the car from killing Herzberg. We will discuss this case in much more detail in Chapter 5, but we mention it here because Vasquez thought the car drove safely on its own ('too much trust'), so she chose to look away from the road for an extended period of time and didn't take the same precautions she might have in other circumstances.[19]

It's important to acknowledge that requiring human operators to remain perfectly vigilant while a car drives autonomously seems to undermine much of what makes autonomous cars attractive. If autonomous cars won't give people time or energy back during their commute, many humans will conclude that they might as well drive the cars themselves. Given the goal of autonomous cars, it should not be surprising that they make their operators less vigilant.

For such reasons, some in the automated vehicle industry – like Google's spinout, Waymo – have argued that it is non-sensical to expect humans to be sufficiently vigilant or trained to override an AI's mistakes in emergency situations. Instead, they opine, we should be making automated vehicles that attempt to remove human actions and decision-making from the loop as much as possible. In kind, hundreds of companies are dedicated to making fully autonomous aeroplanes, drones, cars, trucks, boats, and even flying cars. While these trans-portation modes get around some human vigilance and error problems, their inability to be overridden means their accur-acy and reliability must meet an even higher standard. It's not clear yet whether automated systems will be able to meet that standard outside of very limited driving contexts. In the words of an executive from Lyft, an American rideshare com-pany, executive, 'We will very safely be able to deploy [fully autonomous] cars, but they won't be able to go that many places.'[20] Of course, humans might nonetheless try to take their autonomous cars outside these safe contexts. Adding yet another concern, there is at least preliminary evidence that the mistakes fully automated vehicles will make will disproportionately impact people with dark skin, people wearing long dresses or robes, or people in wheelchairs or mobility scooters ('harm bias'), because the data sets used to train the systems often do not have enough data points from these groups of people for the system to identify them reliably.[21]

Historically, the transportation industry had many frameworks in place to ensure safety, and especially the aviation industry followed these frameworks doggedly. However, these frameworks have proven insufficient for AI transportation for at least two reasons. The first is that transportation AI is still new enough that it is almost impossible to anticipate, and make contingency plans for, all the unique ways it might fail. The second is that transportation regulation and safety industries often lack the financing or expertise to evaluate AI systems adequately. Many of the best transportation safety engineers were never trained specifically in how to assess AI-powered taxis, let alone AI-powered jetpacks, flying cars, or delivery robots. One strategy to manage this acknowledged problem has been to delegate most of the safety certification back to the technology creators themselves. But, of course, technology creators' ability to complete that certification objectively is challenged by both competing financial incentives and the myopic vision creators can have when assessing their own technology. Both problems were cited as reasons why MCAS passed inspections, even though test pilots had repeatedly raised safety concerns.[22]

Another concern with AI-powered transportation is that the more AI-dependent it is, the more likely it is to get hacked ('empowering evil'). For example, in 2017, a well-intentioned 'white-hat' hacker proved that he could gain access to the entire fleet of Tesla cars and not only find their location but 'summon them', a feature Tesla had created to allow drivers to remotely move their cars without anyone in them to help get the cars out of tight spaces. Thankfully, the hacker told Tesla exactly how he achieved the hack, and this problem has since been fixed.[23] Still, it's easy to imagine how malicious actors could use similar mechanisms to intentionally run a car with a targeted passenger off a cliff or into a crowd of pedestrians. Related safety issues result from internet connectivity. When vehicles are connected to the internet to receive software updates (as has been required by Tesla vehicles), they are potentially subject to the same types of hacking attacks as other computers. If cars are accessed by malicious parties, drivers may find themselves unable to start their cars until they pay a ransom or, much worse, find themselves in a car in motion that won't brake or shut off until a ransom is paid.

AI also makes possible an entirely new type of attack which targets the AI itself. These are called 'adversarial attacks' ('new types of harm'). Adversarial attacks are a class of malicious act that includes attempts to fool an AI by providing deceptive input and co-opting or modifying an AI without the user knowing, so that, for example, its output differs from what was intended. One of the most famous adversarial attacks in the transportation sector occurred when a security team showed that they could trick a Tesla Model S into switching lanes into hypothetical oncoming traffic simply by placing three stickers on the road.[24] The car's AI vision system interpreted the stickers as meaning the lane was veering left and failed to detect that the incorrectly interpreted lane was directed towards another real lane, so the AI control system steered the car in the direction of the incorrectly interpreted lane. So far, most documented adversarial attacks on automated transportation systems have been proofs of concept created by well-intentioned entities who are trying to find AI's vulnerabilities so they can fix them, but it is widely acknowledged that it's only a matter of time before enough people gain the skills to implement such attacks in earnest. It seems likely that the field will be in a perpetual arms race between those seeking to execute adversarial attacks and those trying to make the systems not vulnerable to them.

To wrap up, AI-driven transportation definitely has the potential to prevent more deaths and accidents than human-controlled transportation. Nonetheless, the long and diverse list of issues above needs to be satisfactorily addressed for AI's net impact on the safety of the transportation industry to be widely acknowledged as positive.

AI IN THE MILITARY

Military organizations were some of the first to fund and adopt AI technology, and military uses of AI often evoke strong emotional reactions, sometimes in opposing directions. Most people are inspired by reports of AI-driven robots instead of humans carrying messages across mine-laden battlefields,[25] AI-driven autonomous vehicles refuelling planes in the air so human pilots don't have to land and refuel in enemy territory,[26] and AI-guided weapons reducing unintended and unanticipated casualties through improved precision.[27] At the same time, the prospect of AI-driven drones empowering terrorists to kill civilians more easily and cheaply than was previously possible without AI is deeply unsettling ('empowering evil'). AI applications change the safety landscape of military operations in diverse ways. Let's see how some of the safety issues we outlined earlier contribute to those changes.

The Patriot system was the first US military system to employ 'lethal autonomy', or the ability to apply lethal force with little direct human oversight. Created in the 1980s, the system fires missiles to intercept incoming enemy missiles or aircraft. In autonomous mode, the system continuously screens for dangerous targets and fires missiles to intercept them automatically when one is detected. In semi-autonomous modes, humans must perform certain actions before an intercepting missile is launched. The autonomous mode was advertised to be a great success when it was deployed for the first time during Operation Desert Storm in the early 1990s, so the US military confidently used the autonomous mode again during Operation Iraqi Freedom in 2003. Unfortunately, this time the system made tragic mistakes. First, it incorrectly classified an allied British plane as an incoming missile that needed to be stopped ('AI mistakes'). Despite it being used in autonomous mode, the system's operators had a window of about one minute when they could override the system's plan to attack. Due to a collection of contributing factors, they decided to trust the system and allow it to proceed ('too much trust'). The British plane was shot down and its pilots were killed.[28]

In response to the accident, the US military ordered that all Patriot systems should henceforth default to standby mode. Standby mode kept the system operationally ready to act autonomously when quick defence was necessary but required a human to deliberately return the system to autonomous mode before any kind of autonomous attack could be engaged. This extra step of human oversight was supposed to ensure the system would not implement actions that weren't wanted. The US and British militaries were so confident that this policy change would prevent friendly fire mistakes that a British commander announced, 'The Americans have made a rapid and prudent re-evaluation of Patriot rules of engagement. I can categorically assure my crews that there is no danger of inadvertent engagement.'[29] Of course, we are recounting this story because the commander's confidence turned out to be unwarranted.

About two weeks after the British plane was shot down, another US team found itself faced with a Patriot alert about an incoming ballistic missile.[30] According to the military's reports, the commander in charge of the system expressly did *not* want to attack before securing confirmation that the object the system identified was indeed a missile, but also wanted to be ready to attack as soon as that confirmation was available. Thus, the commander ordered, 'Bring your launchers to ready.' The commander thought this command would result in the system being *available* to launch, not engaged in launching. However, since the system was in autonomous mode and had determined there was an incoming enemy missile, it sent out an intercepting missile as soon as the system was taken out of standby mode ('human mistakes'). The system was wrong: there was no incoming enemy ballistic missile ('AI mistakes'). Nonetheless, since the Patriot is a 'hit-to-kill weapon' which is programmed to find something to 'kill' once it is deployed, it aimed for the nearest object it could find, which happened to be a US plane. Again, the plane was shot down and its pilots were killed.

The Patriot system is still in existence and is now used by many countries. Whether or not the faults of the older versions of the Patriot system have been corrected is a matter of contentious debate.[31] The specific cases described above are two decades old, but they illustrate the types of mistakes even contemporary military AIs can lead to.

As in the transportation industry, mistakes by AIs themselves are only one of many safety concerns military AIs generate. First, human operators may trust the AI too much to be able to effectively intervene when the AI messes up ('too much trust'). Indeed, the task force assigned to investigate the Patriot accidents concluded that 'the operators were trained to trust the system's software' and were expected to treat the system's recommendations with reverence.[32] Second, it is becoming increasingly evident that it's unrealistic to expect human operators, in the short time they have to intervene, to assimilate the information and situational awareness they need in order to evaluate the AI system's output accurately. Sometimes the problem will be engaging the human operators' attention fast enough. In the words of one general, 'How do you establish vigilance at the proper time? Three hours and 59 minutes of boredom followed by one minute of panic.'[33] The human brain is poorly designed for these types of contexts. More frequently, it may simply be impossible for operating teams to get the information they need to evaluate the AI system's recommendation in time. Admittedly, these challenges exist in military contexts even when AI isn't used, but AI systems augment the impact these challenges have on accidental or otherwise avoidable loss of life ('new types of harm'). That is cause for ethical concern.

Even if operating teams are vigilant and have access to the information they need for managing an AI system, they will not be able to monitor or use AI correctly if they don't understand enough about how the AI works. Dr John Hawley, a US Army researcher who was involved in the initial development of the Patriot system as well as subsequent investigations of its mistakes, concluded, 'One of the hard lessons of my 35 years of experience with Patriot is that an automated system in the hands of an inadequately trained crew is a de facto fully automated system.'[34] It's tempting to imagine that this problem is easy to get around. Just train the operators! However, military contexts require resources to be bought and mobilized quickly to respond to unanticipated threats. This urgency makes it very likely that AI technology ends up being controlled by soldiers who are not adequately trained in how the AI will behave. For example, when Russia launched a full-scale invasion of Ukraine in 2022, many Ukrainians with no previous military training suddenly found themselves in the Ukrainian military, and the Ukrainian military found itself in possession of donated weapons that it had never used previously.[35] AI-driven military tools can easily be misused in such situations.

The examples discussed thus far relate to defensive military actions, but some safety concerns become more pressing when AI is used for military offence. AI-driven weapons can autonomously target specific people or objects without any human intervention. The Turkey-backed Libyan government used drones to 'hunt down' Khalifa Haftar-affiliated forces in 2020.[36] In 2021, Israeli forces used AI-driven drone swarms to attack Palestinians firing rockets out of Gaza.[37] Such weapons are increasingly being used across the world and are often touted as empowering more accurate and efficient strikes than ever before. But while AI-driven targeting can be extremely accurate for some purposes, it may be less accurate for other purposes (like differentiating civilians from terrorists), and there is nothing to stop a government or other entity from using AI-driven targeting that is not particularly accurate.[38]

Further, by allowing killings to be carried out remotely, autonomous weapons might facilitate dehumanization of the enemy, a phenomenon that paves the way for military actors to feel justified engaging in actions that are decidedly unethical ('AI-mediated dehumanization').[39] For example, using AI for offensive attacks might make it more likely that militaries will make decisions that *increase* civilian deaths and harm, and strategically deny knowledge of the civilian casualties. Such phenomena are argued to be responsible for the unexpectedly large number of civilian deaths caused by AI-driven offensive drone targeting in the Middle East since 2014, and the US's denial of those civilian casualties.[40]

One might respond that civilian casualties or offensive AI mistakes are unfortunate but justifiable when they protect one's country. Even if you are sympathetic to this view, you should still be alarmed by the ways in which AI mistakes can undermine the effectiveness or health of one's *own* military. To illustrate, leaked reports combined with military memoirs indicate that US pilots were more scared of being hit by autonomous 'ghost missiles' from their *own* side during the two Gulf wars than they were of the enemy they were supposed to be fighting, distracting the pilots from their missions ('human distraction') and causing conflict and distrust between US pilots and missile teams who were defending them.[41] It is now also clear that civilian deaths caused by AI-guided drones can take a great psychological toll on drone pilots, even though they are typically operating the drones thousands of miles away from the damage. Although the AI takes care of the targeting after a missile is released, human operators still typically have to tell the AI whom to target and when to start targeting them, so they feel responsible when civilian deaths result. This psychological toll has caused some of the US's best drone pilots to retire from military service, and it can lead to lifelong depression, post-traumatic stress disorder, and suicide in the worst cases.[42] These effects on military unity and on soldiers' mental health and institutional commitment need to be taken into account when calculating the overall impact of military AI mistakes on protecting one's country.

Beyond these concerns, one of the greatest international fears about the rise of autonomous weapons is that they will inevitably fall into the hands of terrorist organizations ('empowering evil'), helping them to cause more harm than ever before with relatively few financial resources ('new types of harm'). Another fear is that, even outside of war, leaders will use autonomous weapons to kill their opponents, possibly including groups with distinguishing physical characteristics that can be targeted with AI-driven facial or body recognition ('harm bias'). Moreover, some uses of AI in warfare have the potential to make military attacks less costly for an attacker and hence more likely. For instance, it is probably easier to attack when the attack is implemented completely by robots and drones than when you have to put real citizens' 'boots on the ground'. In this indirect way, AI might increase military aggression overall.

Ironically, the best protection against military AI may be other kinds of AI. Recognizing this, countries are racing to build military AI capabilities and resisting efforts to restrict AI in military contexts. As an autonomous-weapons expert stated, 'It's not hard to imagine opposing algorithms responding to each other faster than humans can monitor what's happening . . . If we end up with this warfare going at speeds that we as humans can't control anymore, for me that's a really scary idea.'[43] Out-of-control 'AI–AI combat' between conventional autonomous drones or tanks is an unsettling concern, but if countries feel their national security requires having AI systems that control nuclear warheads, an even worse scenario we have to consider is that the world could end up in an AI–AI-driven nuclear war that humans aren't able to stop.

Clearly, the impact of AI on military warfare is multifaceted. AI can both promote and endanger a military's ability to fulfil its duty to protect a nation's interests and values. If we address AI's safety issues sufficiently, AI's net impact on warfare could be to make it more humane. However, if international militaries continue to rush towards AI adoption without managing AI's safety issues, then AI has the potential to make warfare more terrifying than ever before.

AI IN MEDICINE

There are few domains where the promise of AI is more enticing and inspiring than the field of medicine. Medicine aims to prolong life, alleviate suffering, and optimize patients' chances for happy and productive lives.[44] It would be extremely exciting if AI could advance even one of these goals. Encouragingly, AI has already shown potential to make many aspects of medicine more efficient and effective, from streamlining data entry and diagnosing disease to identifying new potential treatments and monitoring pacemakers.[45] In the words of one AI enthusiast, 'A properly developed and deployed AI . . . will be akin to the cavalry riding in to help beleaguered physicians struggling with unrelenting workloads, high administrative burdens, and a tsunami of new clinical data.'[46] At the same time, precisely because medicine is all about saving and improving lives, it is important to remember that medical care touches some of the most intimate and vulnerable parts of human existence. AI's impact on healthcare can therefore be particularly personal, and its safety implications can feel especially unnerving. Let's examine some of those implications.

AI mistakes in the medical domain are often attributable to two practical challenges. The first is that most of the data sets available to train medical AIs include small numbers of patients from certain populations – sometimes too small for what is needed to train the most advanced and successful AIs. The result is that medical AIs that work well for the population on which they were trained often have notoriously poor performance in other populations with different characteristics, such as age or race. More generally, the AI may be deployed in a way or a context that is different from the ones in which it was trained and, as a result, may perform much worse than expected.

The second is that AI has access only to data that are digitized, and the templates for electronic entry of medical data can miss critical information. As a result, medical AIs often learn incorrect models due to lack of important variables, sometimes causing harmful recommendations. To illustrate, having asthma typically makes pneumonia symptoms worse, but one AI designed to recommend whether a patient should be treated as an inpatient or an outpatient counter-intuitively recommended that patients with both pneumonia and asthma should receive less intensive outpatient care, because the AI learned that such patients were *less* likely to die from pneumonia than patients with pneumonia but no asthma. This pattern really did exist in the data set the AI was trained on. However, the pattern was due to the fact that hospitals in the training data set had admitted all pneumonia patients with a history of asthma directly to intensive care units (ICUs) rather than to more standard care units, and the specialized care the patients received in the ICUs led to an approximately 50 per cent reduction in those patients dying. The electronic health records used to train the AI had no fields that represented the presence or absence of automatic ICU admission, so the AI learned an incorrect model whose recommendations could have had disastrous consequences if followed in settings where pneumonia-with-asthma patients were not automatically admitted to ICUs.[47]

Incorrect recommendations from medical AIs might cause little harm if clinicians override them, but many factors, such as time pressure, insufficient information, or overconfidence in the AI, can prevent clinician intervention from being reliable ('too much trust'). It has been shown that physicians have trouble dismissing inaccurate advice from any source,[48] and are legally incentivized to follow AI recommendations, even when they turn out to be wrong.[49] Perhaps this is why the diagnostic sensitivity of human doctors who were usually the most discriminating when looking at challenging mammograms decreased by 14 per cent when the mammogram images were accompanied by incorrect AI diagnoses,[50] and medical residents' diagnostic accuracy of electrocardiograms decreased from 57 per cent to 48 per cent when they were annotated incorrectly by an automated system.[51] So, clinicians will not always correct AI mistakes, and AI mistakes can negatively impact clinical judgment in at least some contexts.

AI-related mistakes are more problematic if they are accompanied by a reduction of doctors' ability to identify them ('deskilling'). Deterioration of skills in the medical domain can have a particularly pernicious effect because medical AIs are usually trained on human doctor ratings, diagnoses, judgments, or opinions. Reductions in clinician accuracy will, ironically, then cause the AIs to become more inaccurate as well. If doctors simultaneously become less able to identify AI's mistakes, medical AI decision-aids might, counterintuitively, cause a vicious cycle of decline in decision accuracy which becomes increasingly difficult to track and correct.

You might think, well, it's OK if an AI system makes mistakes at first, because it takes time to perfect the models; surely the models won't be implemented until they are thoroughly vetted, right? No, unfortunately. Believe it or not, AI tools can be used in some medical settings without any kind of previous testing as long as they are not actively marketed.[52] As a result, many medical AIs that make critical mistakes are already being used in the world's best medical systems.[53] For example, although hundreds of hospitals in the US used some kind of AI-driven model to help them make medical decisions during the COVID-19 pandemic, very few of the models were approved by the Food and Drug Administration (FDA), and many never received any regulatory review at all.[54] When the models were finally reviewed, researchers found that almost all the AI models used during the COVID-19 pandemic provided biased or incorrect predictions. They concluded that few, if any, of the models were ready for regular clinical use.[55]

Moreover, even if AIs are highly accurate overall, they can still augment bias towards certain groups ('harm bias'). One of the domains where AI is most successful is medical imaging. When appropriate data are available, AIs can come close to replicating human clinical judgment when reading medical scans and can sometimes identify medical issues that are missed by the human eye. These benefits are not evenly distributed, however. One recent study of AIs that predict diagnostic labels from X-ray images demonstrated that the AIs fail to detect real medical issues more frequently in underserved populations. Thus, if the recommendations of the AI system were followed, a disproportionate number of patients from female, Hispanic, and Black communities would not receive the care they need because the system would incorrectly decide that they do not have a condition to treat.[56] Critically, the AI systems were assessed using clinician diagnoses, so even if clinicians' diagnoses were biased (as they have been documented to be), the AIs didn't simply reflect these biases – they *amplified* them.

The potential for AI in medicine is great. At the same time, its mistakes influence life and death, and when AI systems are widely adopted, their mistakes can harm dramatically more people than mistakes made by single doctors or provider teams ('new types of harm'). Medical AI's noble task will be to ensure that its benefits outweigh its harms. Doing so will require dedicated monitoring of the issues discussed above, as well as continued consideration of the ways in which each individual AI system affects the doctors, nurses, hospital administrators, insurers, and patients who use it.

AI IN SOCIAL MEDIA

Most social media platforms use some kind of AI – actually, many layers of different AIs – to determine what content to show you. Social media platforms typically want you to stay engaged with the platform so they can show you ads (which usually pay for the platform) and they aim to collect more data about you to target the ads more effectively (and to do other things as well, but we'll talk about that in Chapter 3, on privacy). The more you like what you see, the more engaged you will be with the platform. How do social media AIs show you what you like? They learn what you enjoy based on what you click on, respond to, share, or 'like' – as well as what people whom the AIs classify as similar to you also like. Social media companies come up with as many incentives as possible for you to respond to other people's posts and create posts to which other people will respond. The result is that the entire social media ecosystem has become extremely effective at spreading content and motivating us to try to spread content across the world at speeds previously never even contemplated before the world became so digitally connected.

The motivating idea behind social media platforms is great: keep people connected, no matter how physically far away they are from each other. Social media succeeds at doing that in ways most of us enjoy. Still, as has now been documented and discussed repeatedly, there are many unintended side effects of AI-driven social media platforms, at least as they have been designed and trained so far.

The first side effect is that false information can be spread at least as rapidly as true information. At one point, Twitter users were 70 per cent more likely to retweet fake news than truthful stories and would retweet them more quickly than truthful stories.[57] This false information can lead to severe safety issues. For example, in 2017 rumours were spread over social media in India that kidnappers were abducting children and harvesting their organs. Nobody knows who started the rumours, and no evidence ever materialized to substantiate them, but they nonetheless directly motivated over 150 extremely violent mass beatings of innocent people across India.[58] The dissemination of false information was largely due to private messaging between individuals over platforms like WhatsApp, but it was also reported and amplified on Facebook,[59] and similar videos continue to appear on YouTube.[60] Neither Indian police nor social media companies were successful at stopping the beatings, because in the words of one policeman, 'The speed at which [the false information] goes, nobody can address it, it is almost the speed of light.'[61] As another example in a different domain, health misinformation spread on social media has been shown to convince people to avoid medical treatments that can save their lives[62] and to give themselves dangerous treatments (like ingesting bleach to kill a contracted virus) which lead to their deaths.[63] Many social media companies have invested considerable resources into removing false information from their platforms, including by using AI, but AI is still a main reason false information spreads faster than true information on most social media in the first place.

Another side effect of AI-driven social media algorithms is that they often create 'echo chambers' where you are most likely to see content that reinforces, and usually amplifies, your views and preferences.[64] The echo chambers are a result of AIs doing their best to keep users engaged with the platform as much as possible. We tend to click on and like posts that we agree with, so the AIs increasingly feed us exactly that content. We are particularly drawn to content that makes us feel like we are connected to some kind of group, and we feel closer to that group when we are collectively demonizing another group. As a result, AIs (along with human users trying to maximize their followings) are led to prioritize posts that ignite negative emotions against other groups, and divisive content is especially likely to 'go viral'.[65]

Why are echo chambers a safety issue? The propensity of social media AIs to create echo chambers leads to the proliferation of groups with extreme political and moral views. It turns out that the more positive reinforcement (through likes, shares, or comments, for example) people receive for politically and morally radicalized views on social media, the more likely they are to make physically violent and often life-threatening statements towards other groups.[66] Multiple studies using natural experiments combined with causal modelling suggest that these AI-incentivized and amplified violent statements motivate physical aggression and hate crimes towards victimized groups in real life.[67] In other words, the more frequent and violent the posts against specific groups are on social media in a specific geographic region, the more violent hate crimes are committed against those groups in that geographic region, and the crimes can be tied back directly to what was planned, encouraged, and amplified in social media groups. Even if the threats never come to fruition, they still cause deep disruption to the lives of those being threatened and to the psychological welfare of the groups being targeted.

We do not mean to vilify social media platforms. We personally benefit from using social media, and many social media companies are admirably trying to address the safety issues their AIs pose. Social media also has an important role to play in addressing some safety issues, like when it is used to help identify survivors in natural disasters. While acknowledging these efforts and roles, we highlighted social media in this chapter because it is a paradigm example of a domain that few would label 'safety-critical' but that, with the help of AI, can cause serious harms to many people.

This, perhaps, is one of the slipperiest overall safety problems of AI. AI's success can be its own downfall. When AI affects so many aspects of life so effectively, it almost always comes with unforeseen consequences that cause harm. As AI technology progresses and gets extended into increasingly diverse domains, it is critical that we plan for

AI's unanticipated impacts on people's safety. If we succeed at this, the resulting system may be an improvement over anything we had before. For example, in principle AI can be used to identify and help disrupt the spread of false information on social media platforms, so that eventually false information will spread less effectively on social media than it does over, say, email. If we don't plan for AI's unanticipated impacts on safety, not only will we be caught unprepared and flat-footed when AI starts causing harms, we will also threaten society's confidence in the technology as a whole, undermining its potential to impact our lives for the better.

Predictably unpredictable

For society to responsibly adopt AI, it is essential that we can trust the AIs we encounter. We cannot and should not trust what is not safe. The past decade of AI innovation has made one thing clear in this regard: regardless of whether AI is being used in traditional 'safety-critical' domains or in domains typically believed to involve less risk of harm, the safety implications of AI are predictably unpredictable.

To reduce these dangers, we need to insist that the safety problems we have discussed in this chapter be analysed and addressed whenever *any* AI product is being made, not just AI products made in traditionally safety-critical domains. It is also important to try to anticipate which safety problems a new type of AI product could introduce, either on its own or by virtue of the way it interacts with humans. The more complex or sophisticated an AI is and the more integrated it is into human society, the harder it will be to anticipate its safety impacts.

Anticipating safety issues becomes particularly challenging – and important – when general-purpose AI systems are deployed. Large language models can be used for a wide variety of purposes that individuals and companies are racing to explore. This range of uses is very exciting but also leads to all kinds of safety concerns. For example, these systems can write computer code, making software development more efficient; but that also allows people with bad intentions to create computer viruses or produce code for other nefarious purposes. They can help people assemble information about a medical diagnosis they just received faster than any other available tool; but they can also give bad medical advice about that diagnosis, and are known to fabricate official-sounding information. Each new and creative use of large language models comes with its own unique safety threats. How do we think about the safety of large language models, given the breadth of ways in which the models could be built into new products? Do we need to think about every possible use separately, or are there good general principles to follow that ensure safety across all the different uses?

Two implications are crucial. First, we should be wary of any public, organizational, or regulatory policies based on assumptions that an AI product doesn't pose safety risks when insufficient empirical evidence is available to warrant those assumptions. Second, almost all AI products – no matter how seemingly trivial – must be continuously monitored for potential adverse impacts on health, property, and environment. It can be challenging to address these implications without inhibiting progress towards all the positive things AI can do for society. If our only solution to the safety problem is to not deploy AI at all, some may say, Well, we'll just take our chances with the unsafe AI systems, then.' How do we obtain the benefits of AI while keeping it safe? We will propose some steps towards this important goal in Chapters 6 and 7.

AI technology has the potential to be used safely and even to help promote different kinds of safety, but its benefits must be weighed carefully against its risks. In the words of Stephen Hawking (with whose thoughts we also began this chapter), 'Our future is a race between the growing power of technology and the wisdom with which we use it.'[68] Wisdom requires being humble and clear-eyed about the magnitude of harm AI can create when integrated into real, messy human life.

CHAPTER 3
Can AI respect privacy?

An anonymous app developer, who asked to go by the name Alberto, had seen advertisements for X-ray glasses in old magazines when he was younger. He thought the idea was fascinating. What if he could make X-ray glasses himself? In 2017, he got his chance. A team from the University of California, Berkeley, published the open-source code for an AI system called pix2pix which uses generative adversarial networks (GANs) to learn mappings from input images to output images. Alberto reportedly said, 'When I found out that GAN networks were able to transform a daytime photo into a night-time one, I realized that it would be possible to transform a dressed photo into a nude one. Eureka. I realized that X-ray glasses are possible! Driven by fun and enthusiasm for that discovery, I did my first tests, obtaining interesting results.'[1] The result was an app called DeepNude. DeepNude allowed users to upload a picture of a clothed woman and quickly be rewarded with a nude picture of that same woman. Alberto took DeepNude down after some bad publicity, but the AI used to make the app is still freely available, and many copycat apps have sprung up in its place. In fact, a content marketing agency[2] posted an article titled '7 Best Deepnude Apps 2023 (50 Nude Generators Ranked)'. In the words of one commentator, 'Why would anyone want to be in possession of a fake nude of a woman – of anyone – if not to violate their privacy in some way?'[3]

If the naked body created by the software really does look like the actual woman's body, this example illustrates dramatically how AI can be used to share information that is intended to be kept private. It also raises concerns about objectification, dignity, and harassment, but here we will explore how AI poses risks to privacy, including direct threats that are somewhat obvious and indirect threats that are harder to anticipate.

What is privacy?

The word 'private' is used in many contexts with different meanings. The variety in these uses makes it difficult to provide a unified definition of privacy, and there is little agreement about what exactly privacy is. Nonetheless, experts have proposed definitions including:

Dictionary: 'the ability to be apart from others and be kept hidden from their observation';[4]

Law: the right 'to be let alone' from interference or intrusion,[5] or 'a buffer that gives us space to develop an identity that is separate from the surveillance, judgment, and values of our society and culture';[6]

Business: 'the right to determine whether, when, how, and to whom, one's personal or organizational information is revealed';[7]

Philosophy: 'one's ownership of his or her physical and mental reality and a moral right to his or her self-determination'.[8]

There are also formal, mathematical definitions of privacy which are used in disciplines such as computer science and statistics. For example, one that is often used these days is *differential privacy*, which states, roughly, that any single person's attributes in the data set have a negligible effect on the system's behaviour. This is to prevent anyone being able to infer things about specific individuals from the system's behaviour.

Despite emphasizing different elements, these definitions all share at least one idea: there are some objects, experiences, and types of information we should be able to keep hidden if we want to, especially when they are inherently sensitive, personal, or make us vulnerable to exploitation or intentional harm. Thus, we have privacy to the extent that we have control over whether others observe us, interfere with us, know certain information about us, or determine our reality. We lack privacy to the extent that we do not control information about these aspects of our life.

AI thereby threatens our right to privacy when it causes us to lose control over information about ourselves. This can happen unintentionally through uses of AI that leak private information. It can also be done intentionally when AI is used to uncover private information without permission or through coercion.

It is important to recognize that using AI intentionally to unearth private information is not always a violation of privacy. AI can be used ethically to identify hidden information about us as long as we retain full voluntary control over whether and when the information is learned and shared. For example, most people are generally enthusiastic about AI technology that finds cancer in chest radiographs, even though the presence of cancer is impossible for most people to discern.[9] That's because the owners of the chests in the radiographs want the cancer to be found and have given permission for doctors to use the AI to determine whether the cancer is there. At the same time, most people would be angry if they found out that, without their knowledge, a consulting company used the same cancer-detecting AI on their chest scans to inform recommendations to potential future employers about how much sick leave job applicants were likely to take. Sharing our information that way would be a violation of our privacy because we have no way of controlling what is shared or whom it is shared with.

Why should we care about privacy?

You might be saying to yourself, 'I am not ashamed of who I am or what I do. Why should I care about privacy if I don't have anything to hide?' One answer is that everybody has some information they should aim to protect and keep private for their own good, like bank account numbers and social security numbers. In addition, others can spread seemingly innocuous information about you in ways that will end up harming you because of the way they choose to use that information or because they misinterpret it. You may not think that your religion or your praise for a book is problematic or something to hide, but if others start sending you death threats because of your religion or because they disapprove of the content of the book, then you will likely want the option to keep your religious affiliation and book preferences to yourself.

Moreover, even if you don't have information that you

want to keep private now, there is still likely to be information you want to keep private (or make private again) at some point in the future. Perhaps you will get a medical diagnosis that you don't want others to know about, or you will want to send notes to a co-worker that you don't want your boss reading, or you will want to take down those pictures you posted of yourself on social media. Further, the more your private information is available, the more power others have to make you conform to their preferences. Every time you censor yourself because you are concerned about the social and professional consequences of sharing your true thoughts, you experience the influence of that power. That power can be used intentionally against you or it can intimidate you away from participating in religious, political, or social activities as you want. In all these ways, invasions of privacy can threaten our ability to express our beliefs freely and to live our lives without practical interference. That is why privacy of information is important and why it is recognized as a human right in the Universal Declaration of Human Rights.

How does AI invade your privacy?

Privacy was an issue long before AI was invented. Still, AI adds new ways to invade privacy and new kinds of threats to privacy. Here are a few.

AI CAN UNCOVER PRIVATE INFORMATION

We will begin with some of the most obvious ways AIs identify and disseminate private information. Recall that most AIs are designed to achieve specific objectives. There is nothing stopping AI creators from choosing objectives that unearth private information directly. For example, most people believe we should be able to control who knows about our sexual preferences. Therefore, AIs designed to predict whether we are gay without asking our permission to know that information directly violate our right to keep that information private. This would be true even if the AIs were not perfectly accurate, and even if the AI's creators claim they only created the AI to call attention to AI's potential privacy dangers (as the creators of such an AI argued).[10]

AIs can also be designed to achieve objectives that do not necessarily violate privacy by themselves but that can easily be leveraged to invade privacy. Facial recognition AI is one of the most obvious examples of this phenomenon. Identifying the name of a person in a picture is not necessarily a privacy invasion. Our names are linked to our pictures in many acceptable settings, like yearbooks. Further, facial recognition AI can be used in ways people appreciate and voluntarily give up aspects of their privacy for. In particular, many people enjoy the convenience of AI automatically tagging friends in pictures posted on social media[11] or allowing them to log in to their phones without having to press any buttons.[12]

Privacy problems arise, however, when facial recognition technology dismantles our control over our private information by uncovering it without our knowledge or approval. This loss of control is why many people feel violated when they learn that facial recognition AI is unknowingly being used to board passengers on planes without boarding passes,[13] to provide surveillance at political gatherings or in high crime areas,[14] or to spot potential cheaters and refuse admission to blacklisted gamblers in casinos.[15] In certain contexts, threats to our informational privacy can also endanger two other aspects of privacy mentioned earlier – our rights to be 'let alone' from interference or intrusion, and to 'develop an identity that is separate from the surveillance, judgment, and values of our society and culture'. To illustrate, citizens of New Orleans reported that their lives became unduly stressful and encumbered when facial recognition AIs combined with violence-prediction AIs were used to require citizens to participate in preventative social intervention programmes.[16] Some citizens who had not yet committed crimes were nonetheless required to – or put under a 'fair amount of pressure' to – participate in these interventions because of the AI's predictions.[17] These citizens certainly did not feel 'let alone'. People also report being intimidated away from participating in protests and religious services when they learn governments use facial recognition to track (and potentially punish) those in attendance.[18] Those who are intimidated away from participation are clearly not free to 'develop an identity that is separate from the surveillance, judgment, and values of our society and culture'.

You might be thinking, don't we give up our right to privacy when we are in public? After all, when you walk on a public road or in a shopping mall, other people can recognize your face, learn your location, and draw conclusions about what kinds of clothes you wear without violating your privacy. Is AI doing anything different? This is an important and contentious topic, but at least in the United States, experts across the political spectrum agree that we give up rights to *some* types of privacy when we are in public, but we definitely don't give up our right to *all* kinds of privacy when we are in public.[19] In *United States v. Katz*, the court held that 'what [a person] seeks to preserve as private, even in an area accessible to the public, may be constitutionally protected'.[20] Later rulings and analysis clarified that we have reasonable expectations to privacy in public spaces when '[we], by [our] conduct, have exhibited an actual (subjective) expectation of privacy', and when '[our] subjective expectation of privacy is one that society is prepared to recognize as reasonable'.[21]

These are obviously pretty vague standards, so unsurprisingly there is much debate over how to apply them to our increasingly technology-driven world, where privacy expectations are constantly evolving. Nonetheless, at least one main consideration heavily influences almost all iterations of the debate: society expects some level of 'practical obscurity' in public spaces, including the privacy protection that is afforded by human physical, cognitive, and resource limitations. What does that mean? We know that individual people will be able to look at our faces when we walk down the street and maybe even recognize us, but we don't expect that every person that walks by us will know our name, address, occupation, immigration status, and criminal history because there is little chance that the natural experiences of most people would give them access to that information. When facial recognition AI automatically attaches this laundry list of information to every face it identifies, it uses computational power and aggregations of information that would never be available to most individual humans. This essentially makes practical obscurity impossible to maintain.[22]

Further, facial recognition AI lends itself to tracking people constantly for long periods of time. The US justice system requires police to get a warrant to attach a 24-hour tracking device to a defendant for a month because 'the likelihood a stranger would observe all those movements is not just remote, it is essentially nil', so these tracking devices threaten our right to 'practical obscurity' in ways that must be justified on a case-by-case basis.[23] If such tracking devices are comparable to widespread facial recognition technology which leads to the constant tracking of our whereabouts for extended periods, then it might be reasonable to consider such facial recognition technology as a privacy violation that requires special justification.

It is also important to appreciate the extent to which different AI technologies can be combined to unearth different types of information that are not available to typical human observers. Consider the evidence that depressed individuals generally move around and socially interact less than individuals who are not suffering from depression. As a result, if facial recognition AI is combined with location tracking, the combined system might be used to identify not only the name of a person in public but also whether that person is clinically depressed. One such system is claimed to achieve a 77 per cent true-positive rate (predictions that someone is depressed are 77 per cent correct) and a 91 per cent true-negative rate (predictions that someone is not depressed are 91 per cent correct).[24] A human would not be able to discern this sensitive medical information about us simply by seeing us in the street.

In all these ways, AI facial recognition and tracking technology profoundly erodes our ability to protect practical obscurity in public spaces.[25] Even if public or legal opinion about whether we have a right to practical obscurity starts to change, it is important to realize that this change will represent a profound reduction in the privacy rights we have had in the past.

AI CAN EMPOWER EXCEPTIONALLY EFFECTIVE PRIVACY SCAMS

The cases we have discussed so far all describe uses of AI that are legal in at least some places. One of the biggest threats AI poses to privacy is that it can also be used illegally and with incredible success to trick us into sharing private information. For example, people can be tricked into sharing their bank account information or social security number with an AI deepfake (bots, movies, recordings, or pictures designed to mimic real humans) that they think is a real representative of their bank or the IRS.

Such so-called 'phishing' attacks trick victims into providing private information by sending them emails, telephone calls, or text messages that impersonate communications from legitimate people. Did you ever get an email saying something like 'Your payment settings for your account are about to expire! Go to this website to update your payment details so you can continue to get services.' If that email sends you to a website set up by an entity trying to steal your private financial information, it is a phishing attack.

Although humans create very effective phishing attacks, AI can be used to create even more convincing ones, at even greater scales. To illustrate, the Singaporean Government's Technology Agency showed that AI products available to anyone (either for purchase or for free) can be used to generate phishing emails tailored to people's personalities and preferences. People were much more likely to be fooled by phishing emails written by those AIs than by phishing emails written by humans.[26] Such AI-enabled scams will become increasingly effective and more difficult to detect or prevent as AI technology progresses.

AI CREATES NEW KINDS OF PRIVACY VULNERABILITIES

Many people are not aware that the way AI works and is trained makes new kinds of privacy violations more likely, even if the goal of an AI, itself, has little impact on privacy. We won't be able to detail all these privacy vulnerabilities here, but some examples will provide a feel for how unexpected and unintuitive these vulnerabilities can be.

One type of these new privacy invasions is a phenomenon called 'model inversion attacks'. Model inversion attacks allow people to retrieve the data an AI model was trained on (or characteristics of those data), even if the raw-form data haven't been made available. Researchers have shown that model inversion attacks can be used to reconstruct an image of a person known to be in the training data of a facial recognition model and, somewhat less successfully, to determine the identity of a blurred picture of someone that was not known to be in the training data of a facial recognition model.[27] The same researchers were also able to use model inversion attacks to determine whether a participant in an original data set used to train an AI cheated on their significant other with 86 per cent accuracy, just by having access to the AI model. The goals of the AI model didn't have anything to do with predicting cheating, but the researchers were nonetheless able to retrieve that private information simply through access to the AI's trained model. They could do this because the model's training data happened to include a public data set that was collected to try to relate Americans' preferences for how they like their steak prepared to answers to questions like 'Do you ever smoke cigarettes?' and 'Have you ever cheated on your significant other?'[28]

This sounds concerning, but how often do people have access to other people's AI models? Quite often, actually. In fact, an entire AI-as-a-service (AIaaS) industry has developed around sharing AI models and related services. Big players in this industry include well-known names like Microsoft Azure Machine Learning Studio, Amazon Web Services Machine Learning, IBM Watson Machine Learning, Google Cloud Machine Learning Engine, and BigML. The idea behind AIaaS is that people and entities should have access to the benefits of AI, even if they don't have deep technical expertise or extensive resources. AIaaS entities train AI models that are commonly useful, like facial recognition, text completion, or chatbot systems. Then they let others pay for access to those models, or to have the models applied to their own data. In some cases, AIaaS entities even let other entities use their resources to generate their own iterations of AIaaS services. For example, AIaaS Company A might buy access to AIaaS Company B's emotion recognition system so that Company A can integrate emotion prediction into a different AIaaS system which predicts how happy employees are in their jobs. This means that entities gain various levels of access to AI models on a regular basis, and it can be difficult to track who ultimately has access to AI models or how the models are ultimately used.

Admittedly, model inversion attacks require access to AI models and significant expertise to implement. A second type of new privacy vulnerability caused by AI requires neither of these things. This vulnerability is known as AI models' 'inability to forget'. This phenomenon occurs because sometimes the best way for AI models to make accurate predictions is to 'memorize' aspects of their training data, especially if those aspects occur relatively infrequently.[29] As a result, if prompted the right way, the AI can be fooled into telling you what it memorized, even if you have no access to the model itself. Consider AIs designed to auto-complete sentences in your text messages or emails. Imagine that one of these AIs was trained on the publicly available 'Enron Email Dataset' released by the Federal Energy Regulatory Commission during its investigation of the company, which consists of the full text of several hundred thousand emails sent between Enron employees. It turns out that some employees used emails included in this data set to send private and sensitive information, like social security numbers and credit card numbers. Once this was discovered, two years after the original data set was posted, the social security numbers and credit card numbers were removed from the public data set so that they were no longer easily available.[30] Nonetheless, if you prompt an AI that was trained on the original, unscrubbed version of the Enron email data with a phrase like 'My name is [fill in the name of an Enron employee] and my social security number is. . .' the AI might complete the sentence with the actual social security number of the employee. Alternatively, you could prompt, 'My social security number is. . .' to gain an anonymous number, then follow that up with, 'Since my social security number is [insert the number the model gave at the first prompt], my name is. . .' and the model might return the name of the person associated with the number. In this way, AI makes it very difficult to correct privacy violations or allow people to change their minds about what information they want to share.

Systems such as ChatGPT, trained on enormous amounts of data, show how quickly these concerns can snowball. If the next version of GPT is trained on previous interactions that users had with ChatGPT, what guarantees do those users have that others will not be able to recover information about their interactions? Once you enter a question into ChatGPT, will it remember that question forever (at least in a sense) and make it possible for others to associate it with you? Indeed, Microsoft and Amazon have warned their employees not to share sensitive information with ChatGPT because this information may later appear in its output, so it is widely acknowledged that these kinds of privacy violations are possible and maybe even likely.[31]

Researchers are constantly reporting other ways AI models, and the technical infrastructure surrounding them, cause new privacy threats.[32] The important thing is that these new types of technical vulnerabilities exist, and unlike the phenomena described in the previous two sections, they are usually unintended and unfortunate side effects of using AI, rather than results of AI being designed or used explicitly to uncover private information.

The AI ecosystem incentivizes hoarding and selling of private data

So far, we have focused on AI's direct threats to our privacy. It is equally important to manage the indirect threats AI poses to our privacy through the ecosystem and marketplace that have evolved to take advantage of AI. In the words of one writer, this ecosystem 'is based on collecting as much personal data as possible, storing it indefinitely, and selling it to the highest bidder'.[33] Why?

THE RACE FOR PRIVATE DATA COLLECTION

Entities are competing fiercely to become AI technology leaders and to benefit from the myriad of valuable ways AI might be used in the future. It is well known that the more diverse and vast the data you use to train an AI are, the more accurate the AI is likely to be, and the more things the AI will likely be able to predict successfully.

To understand why this is true, consider the following. I will not be able to discern much about your love life from tracking your movements in isolation (unless I happen to know that the locations you frequent are establishments tailored to groups with certain sexual preferences, or something like that). However, if I track the movements of everybody in the world, I can make pretty good predictions about whom you likely sleep with and how often you sleep with them by looking for overlaps between your location and other people's location during night-time hours. In this way, aggregating data across individuals allows me to make fundamentally novel insights that data about one individual can't provide on its own.

CHAPTER 3

AIs benefit from the same phenomenon. If a person is shopping for skis on one website, as well as for flights to Denver on another website, then knowing about both of these interactions allows an AI to predict that an ad to this person about a particular ski resort near Denver is likely to be successful. As a result, entities are strongly motivated to collect as much data as they can in the hope that AI (or other types of analysis) can eventually be used to monetize the data or turn it into something of value down the line, even if (sometimes especially if) that data is private.[34] As the title of an article in the *Economist* declared, 'The world's most valuable resource is no longer oil, but data'.[35]

How does the lust for more data result in our loss of privacy? One of the most effective ways to collect vast amounts of diverse data is to give individuals free services in return for sharing data, even if the data that are collected have nothing to do with the services provided. When we are one of those individuals, we often happily use the offered platform, app, or device completely unaware that entities are collecting private data from us in return. Our ignorance might be because we were never actually told our data were being collected, or because we gave consent for the data to be collected in an incomprehensible terms-of-use agreement that we never read anyway.

To give some sense of how common it is for our data to be collected without us realizing it, one study found that 19 out of 21 tested phone apps sent personal data to a total of approximately 600 different primary and third-party domains, and some of them did so continuously, even when the app was not in use.[36] In another study, The Haystack Project

94

found that more than 70 per cent of apps assessed were connected to one or more other data trackers, and 15 per cent of the apps were connected to five or more trackers. Twenty-five per cent of the data trackers harvested at least one unique device identifier, such as a phone number. More than 60 per cent of the trackers sent information to servers in countries known to use mass surveillance in their populations, raising questions of whether governments in those countries had access to the shared information as well.[37] Websites share data just as much as applications, if not more so. Approximately nine out of ten websites share user data with an average of nine different third parties, typically without the users' awareness.[38]

Critically, the data these companies, websites, or apps share with third parties often include information we generally want to keep private and assume is being kept private. For example, in 2018 reporters uncovered that the social networking/dating app Grindr shared information about whether its users were HIV-positive with Apptimize (an app-optimizing company) and Localytics (a marketing platform for apps), completely unknown to Grindr users.[39] Further, if you use Gmail, you may not know that you have agreed to allow third parties to read your emails, including any private information you put in those emails. According to the chief technology officer at eDataSource Inc., an email intelligence company, 'Some people might consider that to be a dirty secret . . . [but] it's kind of reality.'[40]

These types of large-scale private data collection and sharing do not necessarily use AI. Still, they are largely motivated by the universal acknowledgement of AI's promise.[41]

As one blog writer put it, 'AI is useless without data, and mastering data is insurmountable without AI.'[42]

THE RACE FOR PRIVATE DATA PREDICTION

Even if entities do not collect or have access to your private information directly, the more data they have, the better their ability to predict your private information accurately, especially if they use AI to make those predictions. Accurate predictions about your private information can be valuable for many different reasons. Retail companies may use these predictions to improve personalized marketing, like when they target baby supply promotions to pregnant women or health equipment promotions to people with diabetes. Companies like Cambridge Analytica make predictions about people's personality types and preferences to help political groups tailor their messages in ways that maximize influence on voting behaviour.[43] Insurance companies may want to predict various aspects of your behaviour or private health information to optimize your health premiums.[44] Financial companies may want to predict whether you are likely to have large hospital bills when they are deciding whether or not to grant you a loan.[45] Pro-life governments may want to predict which citizens are likely to be seeking abortions.[46] AI is increasingly being used to predict this type of private information more accurately from large data sets and can be used much more in the future, especially as AI's accessibility grows.

THE MARKET FOR PRIVATE DATA SELLING AND SHARING

Data, like any valuable resource, aren't just attractive for the way they can be used. They are also attractive for the money they can make when they are sold. An over $200 billion industry of 'data brokers' has developed to buy and aggregate historical data about people which they can sell to other entities.[47] Sometimes the data or profiles data brokers sell include only raw data, and other times they include predictions the data brokers made about you (with varying degrees of accuracy). The data sets often include very private information. For example, the medical data broker MEDbase 200 reportedly sold lists of rape victims, of people with erectile dysfunction, and of people with HIV.[48] Moreover, our information is startlingly cheap to buy. MEDbase reportedly sold its information for $79 per 1,000 names.[49] Thus, it has become fairly easy and affordable for anyone to buy private information about you, and companies and entities regularly take advantage of this opportunity.[50]

Entities that do not formally consider themselves data brokers also frequently collect data about us as part of their financial business models. Just like data brokers, these entities may sell raw data or profiles they have created about us based on the data points they assemble. For example, you might not expect The Weather Channel to be in the finance business, but they have sold the location data their weather app collects to hedge funds to help them value businesses based on the foot traffic The Weather Channel app records at the business premises.[51]

Sometimes entities also trade our data for business services, access to customers, or advertising agreements. Facebook (now

Meta), Twitter, and Tiktok do this regularly, as does PayPal.[52] A lot of this data sharing makes our life easier. For example, did you ever wonder how you can use your Google account to log into a random, unrelated website? That convenience is made possible through private data sharing between business partners. The problem is that much more data may be shared in these agreements than customers realize. To illustrate, you may not realize that when you authorize PayPal to integrate with websites, you allow PayPal to share your shopping history, personalization preferences, pictures, and disability status (among many other things) with those websites.[53] Such practices lead us to misinterpret privacy policies that assert, 'We do not sell your personal data.' Just because an entity doesn't *sell* our data doesn't mean it doesn't *share* our data with a lot of other entities we know little about.

In sum, the promise of AI is at least partially to blame for a vast ecosystem where a huge variety of information about us is easily collected, predicted, and combined, often without our knowledge. The more of our information that is available, the more AI can be used intentionally and unintentionally to learn private information about us, and enable others to leverage that information to manipulate, coerce, or scare us into actions that might cause us personal harm.

How can we protect our privacy from AI threats?

It should be clear by now that AI and the associated ecosystem inspire many new threats to our privacy. Can we do anything about it? Let's examine the most common strategies people consider.

REQUIRING CONSENT

Privacy is about *control* of information, not the permanent hiding of information. If a woman is sexually assaulted, that information is private to the extent that she can control who finds out what happened and how many details they find out. She should not be coerced or tricked into sharing what happened, but she can share what happened if she wants to, and she gets to choose how much of what happened is shared. Acknowledging this, it seems that a straightforward way to protect private information from AI or AI ecosystem threats is to require that we give consent before any of our information is collected or shared. This step should enable us to retain control. We can consent if we want to, refrain from giving consent if we want to, and generally be in charge of whether and when others learn information about us. Ah, if only life were so simple!

'Terms of use' and 'privacy policy' agreements are supposed to describe how our data will be used. Therefore, when we sign them, it is assumed that we approve of the way entities named in the agreements will use our data and that such uses do not violate our privacy rights. However, there are obvious problems with these ideas.

First, we have all clicked buttons on consent forms or terms-of-use screens without reading what we are consenting to in order to save time. After all, most privacy policies take eight to twelve minutes to read[54] and use language that even experts do not understand,[55] so it should be no surprise that most of us do not read them. One study found that 74 per cent of participants skipped reading a social media company's fictional privacy policy or terms-of-use agreement, and those who did read any of it only read for about a minute, even though it should take about 30 minutes to read the privacy policy and 16 minutes to read the terms-of-use agreement. As a result, over 93 per cent of participants agreed to both policies presented in the study, despite the policies indicating that consenting participants would have their data shared with the National Security Agency (a United States intelligence organization) and would donate their firstborn child to the social media company.[56] This study makes it clear that signing a consent form does not mean that participants have actually read it.

At the same time, it seems unreasonable to expect people to read most terms-of-use policies for products they consider, given how long they would need to read. Remember that if an app or platform reports sharing data with third parties, we also have to read the terms-of-use policies of all those third parties (and the parties they, in turn, share data with) to fully understand what will happen to our data. As a result, it has been estimated that it would take Americans approximately 244 hours per year to read all the privacy policies of the digital products they encounter.[57] This is not a realistic time commitment to require of most consumers.

Even if people did read the terms-of-use policies, it is unlikely they would understand them. As noted above, many policies use legal and technical terms that are incomprehensible to the average user.[58] Moreover, the data sharing, selling, and prediction ecosystem has become complex enough that most people will not have sufficient background knowledge to understand the implications of data sharing policies. Adding to all this, many entities intentionally use 'dark design' to create interfaces that explicitly exploit human cognitive biases to make users more likely to give their consent to whatever is asked.[59] We are responsible for our own choices to skip consent forms. However, consent does not really protect privacy if people are not realistically able to learn the implications of their choices or are manipulated into consenting.

There is also a third reason to question whether consent via terms-of-use agreements protects privacy. This reason is potentially more controversial. To have full control over how your data are used, you must be able to choose to keep your data private without incurring disproportionately large costs. It is therefore concerning that these costs are becoming increasingly high. For example, schools often require children to complete assignments using web browsers or apps on tablets or computers, and students might fail those assignments if they are not completed; but those web browsers and apps track and share what the children look up and how they perform on the assignments.[60] Employees are often required to use employer-supplied Gmail accounts or use biometric security systems, and must do so in order to remain employed, even though Google has allowed third parties to read its emails[61] and those security companies mine our biometrics for personal information.[62] Daycares require us to download apps to sign our children in and out in order to access their services, even though those apps share pictures and information about our children with other parties.[63] These everyday necessities raise serious doubts about whether people truly have free choice to reject all data tracking or privacy-threatening technology. Consent does not really protect privacy if people are coerced into consenting out of rational fear of the real consequences of not consenting.

In sum, requiring informed and free consent for information sharing is an important tool for protecting privacy. Unfortunately, it won't save privacy on its own.

ANONYMIZING DATA

Can't we just anonymize data when they are shared? Wouldn't that solve the problem? Unfortunately not.

First, let's clarify what anonymization is. Anonymizing data removes or replaces any information that directly identifies the individual person whom the data represent. People's names might be replaced with random number strings or people's addresses might be removed altogether (so that, for example, somebody can't just use publicly available websites to look up who lives at the addresses in a data set and then figure out whom the rest of the data in the set are associated with, even if the data set itself does not contain people's names). This is a good start to protecting the privacy of people in the data set, but it is by no means sufficient in today's data ecosystem. The critical issue is that combining data sets makes it easy to re-identify people in supposedly anonymous data sets, and AI makes re-identification even more accurate and more efficient.

Here is the general idea. Imagine a data set with columns for customer ID, ZIP code, birth date, and gender. The customer ID is a random number, so there is no way to use it to figure out whom it represents. However, one of the customers lives in a ZIP code with a small population, and it turns out they are the only male in that ZIP code born on their birthday.

As a result, the values in the ZIP code, birth date, and gender columns allow that customer to be 're-identified' in the data set, even if that customer's name is not explicitly included in the data set. This won't happen just for people in small towns. About 87 per cent of the population of the United States can be uniquely identified through a combination of only their birth date, gender, and five-digit ZIP code.[64] Of course, to re-identify people, you do need help, either from people in the town who know their neighbours well or through voter registration records from the town, which you can buy for $20.[65] But that's where the impact of the data sharing economy we talked about earlier comes in. It is now very easy to access the kind of additional information needed to re-identify someone, either through the data they provide publicly (such as through LinkedIn) or through the vast data sharing and buying ecosystem. 99.98 per cent of Americans can be correctly re-identified in any data set using fifteen demographic attributes.[66] Fifteen attributes might seem like a large number of variables to collect, but recall that most data brokers have thousands of data points on each individual, and most companies have at least hundreds. Thus, if data brokers or companies buy or are given access to a supposedly anonymous data set, they will almost certainly be able to re-identify many of the people in that data set.

As another researcher in the field said, 'It's convenient to pretend it's hard to re-identify people, but it's easy. The kinds of things we [do to re-identify people in supposedly anonymous data sets] are the kinds of things that any first-year data science student could do.'[67] Given that hospitals and health providers are now sharing 'anonymous' data with

Google, Apple, Microsoft, Amazon, and IBM, companies that likely already have hundreds of demographic variables about you, it is reasonable to assume that most anonymous healthcare data can be re-identified by parties it is sold to, despite privacy laws and regulations.[68]

Data sets with many time points, such as purchase transaction histories or time stamps of social media posts, are particularly susceptible to re-identification. To illustrate, researchers were able to re-identify Twitter users in a group of 10,000 users with 96.7 per cent accuracy from seemingly anonymous and innocuous 'metadata' about their posts, such as when the post was made, the number of people the poster was currently following, and the number of total tweets made by the poster.[69] Importantly, much of this metadata was accessible to any developer who wanted to integrate their services with Twitter through application programming interfaces (APIs) that allow apps and websites to talk to one another in order to create more seamless user experiences.

With the help of AI, anonymous geolocation data can be even easier to re-identify. AI experts demonstrated that 95 per cent of the individuals in a large data set with hourly GPS data could be identified using only four GPS spatio-temporal points.[70] Thus, it is not that hard for AI to identify you in a crowd and track your movements with a relatively small amount of GPS data.

This power is concerning, given reports that almost half of Android apps and a quarter of Apple apps use your phone's GPS function to record exactly where you are at any given time, and share that information with many third parties or even make it public.[71] In one case, Polar (a fitness app used by many in the military) made it easy for its customers to share their exercise routes on social media. By looking at exercise routes executed around locations such as the National Security Agency headquarters, secret service bases, and people's home residences, journalists were easily able to cross-reference Polar's public data with public social media posts to identify almost 6,500 military officers' identities across 69 countries, including many who worked in places with national security implications (like military stations near the North Korean border). They were able to ascertain not only where these people lived, but also what their travel routes to work were.[72] This kind of information could clearly be used in dangerous ways by extremists or state intelligence services, especially given that some of the data was about personnel who worked at nuclear weapons storage sites. AI can unearth this kind of information from our supposedly anonymous geolocation data even faster and more effectively than the journalists did.

In sum, in the words of journalist Karl Bode, 'It's not clear how many studies like this we need before we stop using "anonymized" as some kind of magic word in privacy circles.'[73] At best, data are very, very hard to anonymize and are usually not handled appropriately to ensure anonymity. At worst, there is so much data available in the world that it is no longer possible to truly anonymize a data set.

NEW TECHNOLOGIES

AI technology has caused many new privacy issues, but can technology – AI technology, in particular – also be used to help address some of these privacy issues? This time we can offer some good news – yes, it can!

One potential way to ensure that someone's individual information can't be retrieved from a set of statistical analyses is to blur everybody's identifying data in such a way that analysis results remain the same, regardless of whether a particular individual's data are included in the analysis. This concept, already mentioned at the beginning of the chapter, is called 'differential privacy'.[74] A related method is to use synthetic data sets that have the same properties as real data but include fake details.[75] Another promising approach, called 'homomorphic encryption', allows AI algorithms to be trained on encrypted data instead of on non-encrypted data.[76] AIs can also be designed to identify deepfakes or phishing attacks and send them to our spam folders and attach warnings to them.

These are just some of the ways that privacy protection can be built into AI ecosystems or algorithms. Researchers and practitioners are working hard to develop others, and we are optimistic some will be successful. No one privacy-enhancing technology can protect all forms of our privacy, but, especially if multiple privacy-enhancing technologies are used together, possibly supported by laws that require their use, our privacy will at least have a fighting chance.

Do we value privacy?

We have tried to give reason for optimism in the section above. At the same time, it's important to clarify that we don't think technology is sufficient for solving all the privacy issues that AI and the AI ecosystem pose. There are many social, contextual, and practical issues that will have to be addressed as well. We will talk about some of these issues in broad strokes in Chapter 7. However, we want to address one critical social issue here: the 'privacy paradox'.[77]

Despite claims by marketers or data collectors that people believe sharing their data is a fair trade for personalized services,[78] a survey by the Pew Research Center determined that 81 per cent of the public believe that the risks they face because of data collection by companies outweigh the benefits and that they have little to no control over how their data will be used by companies.[79] In other words, at least for the time being, people still value privacy. Here's the paradox, though. Most of us are not willing to stop using apps or technologies that share our data without our consent, even if we are annoyed and disturbed by the fact that our data are shared. This well-documented pattern has been called 'privacy resignation'[80] or 'privacy cynicism'.[81] It may be due to at least two concerns.

First, many of us justifiably conclude that, in today's world, our desire for privacy conflicts with other critical life needs and obligations, and we need to prioritize those other life needs and obligations. For example, we may hate the fact that AI-based hiring platforms predict our private information and share that information with other parties, but we still need to apply for jobs that require the use of those platforms, because we need to maximize our chances of securing income. Similarly, even if there is an alternative social media platform that we would much prefer to be on due to its better privacy protections, that is of little value to us if all of our friends and family are on a different one.

Second, it is easy to assume that any efforts to protect our privacy are futile. After learning or personally experiencing the types of phenomena described in this chapter, we may feel the forces taking away our control over our data are too great for us to do anything about.

The privacy paradox could be AI's greatest threat to privacy. AI's promise has contributed to a cultural ecosystem where our privacy is violated so continuously that many in society no longer do much to try to stop it. This trend poses a grave danger. Privacy is worth protecting. It enables us to maintain our autonomy, individuality, creativity, and social relationships by preventing exploitation and affording critical psychological and functional benefits. Despite the way privacy violations have been normalized and naturalized, privacy violations are not an inevitable consequence of AI. We do not have to accept a culture that pits privacy against innovation and pursuit of knowledge. AI and privacy can co-exist in a society that fosters innovation as well as human dignity and autonomy. But we need to be willing to work hard to make sure that all of these values are sufficiently respected.

Can AI be fair?

British students are required to take A-Level exams to get into university, and their exam scores dramatically impact which universities they are admitted to. Due to social distancing regulations introduced during the COVID-19 pandemic, many British students were not able to take their A-Level exams in person in 2020; so, in lieu of traditional exam scores, the UK government decided that an algorithm would be used to award each student's exam grades. The algorithm awarded grades based on teachers' assessments, individuals' performance on practice exams, and, most critically for our current discussion, schools' exam track records in previous years. The latter was used to 'correct' for grade inflation and standardize results across the country. The result? More than 40 per cent of students ended up receiving lower exam grades than they or their teachers expected, causing an outcry across the country. Worse, grade demotions were applied to students from less affluent areas most frequently due to their schools' overall past exam performance, and many affected students had their acceptances at universities revoked due to their lower-than-expected exam grades.[1] According to one report, 'Bright students in historically low-achieving schools were tumbling, sometimes in great, cliff-edge drops of two or three grades, because of institutional records they had nothing to do with.'[2]

This case is not exceptional in the AI world. Headlines frequently suggest that AI is unfair to disadvantaged groups in various ways.[3] AIs commonly used for hiring, firing, promotion, home loans, and business loans often disfavour Black, female, immigrant, poor, disabled, and neurodiverse applicants, among other groups. AIs that detect skin cancer work less well on darker skin, so doctors who rely on them will be less likely to treat people with darker skin in time.[4] And, as we will see, AIs that predict crime recommend denying pretrial release and bail more often to defendants who come from underprivileged groups. All of these performance patterns are biased in the sense that good or bad consequences are awarded disproportionately to certain groups of people, usually in the form of harms to already-disadvantaged groups and benefits to already-privileged groups. When such biases are unjustified, as they usually are, they are considered to be unfair or unjust – terms that we will use interchangeably.

But if AI is so 'intelligent', shouldn't it know better than to be biased? For all the many surprising advances that AI technology makes, this is one of the arenas where it continues to struggle. One of the most fundamental reasons machine learning, in particular, is frequently biased is that it is very difficult (and often expensive) to assemble data sets that have all demographic groups and interests represented equally, and trained models are usually more accurate at making predictions about groups that are well represented in its training data than groups that are not. This is why we repeatedly see phenomena like face recognition AIs being less accurate at identifying dark faces than light faces – many facial training data sets predominately comprise Caucasian faces.[5]

A more general reason AIs end up biased is that humans and human social structures are often biased, and our biases are readily built into the AIs we design and create. Every time a human decides what data to collect, labels a data point, decides what information should be fed into an AI algorithm, chooses a goal for an AI to pursue, decides how to evaluate an AI model's performance, or decides how to respond to an AI prediction, opportunities are created for our own human biases to be reflected in an AI. In many ways, AIs end up being reflections of their human creators and the context their human creators inhabit. As some say, 'bias in, bias out'.

These two overarching causes for AI bias are so pervasive and challenging that most experts, regardless of their level of technologic optimism, agree that AI systems (like humans) are almost never perfectly just or fair. This raises the critical questions: should we use AI when we know that it can contribute to injustice? And is there perhaps some hope of designing AI systems that would actually reduce injustice, perhaps even in settings where AI currently does not play any role? We will explore these questions in this chapter, beginning with a few brief words about the nature and kinds of justice.

What is justice?

At least since Aristotle,[6] philosophers have distinguished several areas or kinds of justice or fairness that apply to families, society, and the law.

Distributive justice concerns how burdens and benefits are distributed among individuals and groups. It seems unfair or unjust for businesses to refuse to hire applicants from a disfavoured group, for municipalities to provide better schools or more police protection to a favoured group, or for countries to require or allow only some groups and not others to serve in the military. Such practices might be reasonable in certain circumstances, but justifying such inequality would take at least some special reason.

Retributive justice, in contrast, concerns whether a punishment fits the crime, or, more generally, whether people get what they deserve. Punishments can be unfair by being too harsh or too lenient. It seems unfair to sentence a car thief to life in prison, because that punishment is too harsh for that crime. On the other hand, it also seems unfair to sentence a rapist to only one day in jail, because that minor punishment is too lenient for such a horrible offence.

Procedural justice differs from both distributive and retributive justice. It concerns whether the processes or procedures used to reach decisions about how to distribute benefits and burdens are fair. Even a murderer who confesses and is clearly guilty still deserves a fair trial. Similarly, a procedure for selecting political leaders would be unfair if certain races or genders were denied the right to vote, even if the same candidates would win anyway.

In sum, different cases can be unfair or unjust in different ways. There are more kinds of justice,[7] but these three kinds of justice – distributive, retributive, and procedural – are enough to keep us busy. We will explore issues of AI fairness that touch on all three types of justice through focusing on uses of AI in the legal system.

Who goes to jail before their trial?

The police make over 7 million arrests every year in the US.[8] After arrest and booking comes an arraignment, where a criminal defendant appears in court to hear the charges against them and submit a plea. This arraignment is typically combined with a bail hearing, in which a judge decides where the defendant will live while waiting for the next hearing or trial. The judge can decide to let the defendant go home (or wherever they want) with only a written promise that they will return at the next required court date. The judge can also require the defendant to stay in jail during that time if they think the defendant is likely to fail to show for their court appointment or commit a crime in the meantime. An intermediate option is to allow the defendant to go home until their next required court appearance if, and only if, they pay a certain amount of money as a security deposit to help ensure they will return for their scheduled court dates. The security deposit is returned to the defendant if they appear in court when requested, but it is kept by the court system if the defendant misses any of their scheduled court dates. The financial security deposit is sometimes called 'bail', but the term 'bail' can also refer more broadly to any pre-trial release with a promise to appear at the trial, even if no security deposit is required. We will refer to the decision of where a defendant should reside under which conditions while waiting for trial as a 'bail decision'.

Bail decisions have serious consequences for many people. Defendants who are granted and can afford bail are allowed to go home to their families, friends, and jobs while awaiting their trial, which positively impacts their mental and financial health, as well as the mental and financial health of their families. In contrast, defendants who are denied any opportunity for bail are forced to remain in jail until their trial, perhaps for months. This detention imposes personal costs that include not only loss of physical freedom but also often loss of their jobs and loss of their homes if they cannot work or pay the rent while in jail. In these ways, and more, they and their loved ones end up paying dearly, even if they are eventually found not guilty.

Despite these heavy costs of denying bail, courts need the option of denying bail. When a defendant released on bail skips town or threatens or murders witnesses, for example, most people wish that defendant's bail had been denied. On the other hand, when there is little risk that a defendant will commit additional crimes while out on bail, and it is very likely that the defendant will show up for their trial and spend their time out on bail productively (such as by working to support their family), then the costs of denying the defendant bail seem too harsh.

Importantly, judges in the United States are not supposed to make these bail decisions on the basis of whether they think the defendant is guilty. Assessments of guilt come later, during the trial. Instead, judges are typically supposed to base their bail decisions solely on two predictions of what the defendant will do if released: will this defendant flee and fail to appear at the trial? Will this defendant commit another crime while out on bail?

The problem, of course, is that nobody can know for certain in advance what an individual defendant will do while released on bail. Nonetheless, judges need to decide where a defendant will reside while waiting for trial, and these decisions usually need to be made very quickly. The average arraignment in New York City is estimated to last only six minutes because of the large caseload and the small number of arraignment judges. The time pressure makes it unrealistic for judges to ponder or even familiarize themselves with all the relevant details of each case. The time pressure may also make it more likely that judges will rely on some of their documented implicit bias towards or against certain groups when making decisions.[9] Thus, courtrooms across the United States have turned to AI for assistance because they believe that AI can make more accurate predictions from complex information and show less bias than humans.[10] Let's consider both of those important claims.

Human judges versus AI: who is more accurate?

Judges themselves admit that they make many mistakes when predicting future crimes. The American Law Institute (ALI), whose members are renowned judges, lawyers, and law professors elected by their peers, wrote this about sentencing:

> Responsible actors in every sentencing system – from prosecutors to judges to parole officials – make daily judgments about . . . the risks of recidivism posed by offenders. These judgments, pervasive as they are, are notoriously imperfect. They often derive from the

intuitions and abilities of individual decisionmakers, who typically lack professional training in the sciences of human behavior . . . Actuarial – or statistical – predictions of risk, derived from objective criteria, have been found superior to clinical predictions built on the professional training, experience, and judgment of the persons making predictions.[11]

These eminent legal experts believe that statistics are better than judges at predicting risk of future crime, and AI predictions are based on statistics. Even so, this admission is about sentencing rather than bail decisions, so what about bail?

To see whether humans or AIs are more accurate in predicting bail violations, we need some way to determine which predictions are correct. That is difficult to do given that we can never know whether defendants who were kept in jail before their trials would have fled or committed a crime if they had been released. Still, we can try to get some idea of how good judges are at predicting bail violations by examining past legal records to determine which defendants who received bail failed to appear for trial or committed crimes while out on bail, and by comparing an AI's assessment of those same defendants.

In one study looking at bail decisions in New York City,[12] the defendants whom an AI (under our broad definition) classified as risky failed to appear for trial 56 per cent of the time, committed other new crimes 63 per cent of the time, and even committed the most serious crimes (murder, rape, and robbery) 5 per cent of the time – all much more than defendants whom the AI did not classify as risky. It is important to note here that these statistics themselves may be biased. For example, we don't actually really know how many defendants committed new crimes. All we know is how many of them were *caught* committing a new crime, and if police disproportionately look for crimes in one population, they are likely to catch disproportionately many crimes in that population. Also, more people in such a population may be convicted of a crime they actually did not commit, but the fact that the person is actually innocent of course won't be in the data fed to an AI. Even keeping this very important caveat in mind, some of the defendants the AI identified as high risk seem to pose real dangers. Despite these dangers, human judges released 48 per cent of the defendants whom the statistical predictor rated as the riskiest 1 per cent. In short, the judges treated many high-risk defendants who violated bail as if they were low risk. Conversely, judges also denied bail (or required bail payments that defendants could not afford) to many innocent defendants whom the statistical predictor rated as low risk. So the AI seems to correctly classify some cases that human judges misclassify.

Judges might respond that they base their bail decisions on important features of defendants that are left out of tracked statistics. For example, a judge might deny bail to a defendant who has a gang tattoo that the judge sees as highly predictive of crime risk, even if that tattoo is not in the database used to train an AI. Judges also might use family presence in the courtroom as a signal the defendant has social support that will reduce the likelihood of recidivism, even if most bail AIs do not take family support into account. However, we don't know whether judges really use these kinds of 'untracked' features in most cases, and such features don't account for all of judges' prediction mistakes, especially those that lead them to release high-risk offenders, including some with gang tattoos and without family at the trial. A more parsimonious explanation of human judges' failures to anticipate bail violators accurately is that many judges may base their decisions on factors that do not actually support their predictions.

It would be understandable if this were the case. Lots of behavioural science shows that even the best-intentioned and best-trained human decision-makers are often misled by information they mistakenly think is important.[13] So even if judges are more accurate at predicting bail violations than most other people, it wouldn't be surprising at all if they were still frequently misled. The question is: can AI become more accurate than human judges by processing much more data much more quickly than judges?

At least to some extent, the answer seems to be yes. The results described earlier were from bail decisions only in New York City, but similar results were found when analysing bail decisions in a national data set covering over 151,461 felony defendants arrested between 1990 and 2009 in 40 large urban counties across the US. According to that replication, basing bail decisions on AI could either reduce the crime rate by 19 per cent (holding the release rate constant) or reduce the jail rate by 24 per cent (holding the crime rate constant). Other studies have shown that criminal recidivism algorithms have decent accuracy as well.[14] That seems like good news.

Admittedly, we should never just assume that an AI will be more accurate than humans, and even if an AI is more accurate in some contexts, that doesn't mean it will be more accurate than humans in other contexts or at other time points when new factors become important.[15] Big problems will also arise if humans trust AIs more than their accuracy performance deserves. Still, if similar accuracy to the studies above can be achieved in other data sets and in different types of settings across the country, AI might sometimes be a useful tool for making more accurate bail decisions.

Making fairness explicit

Even if the accuracy of AI prediction tools in bail decisions is promising, another worry is that an AI's predictions could have disproportionately harmful effects on disadvantaged groups. In discussing this issue, we will focus on whether AI tools that predict criminal recidivism are fair across racial groups. For example, the COMPAS (Correctional Offender Management Profiling for Alternative Sanctions) criminal risk assessment tool was designed to aid judges in deciding which defendants should get alternative sanctions, such as drug treatment programmes or rehabilitation services. It has also come to be used in other parts of the criminal justice system, including bail, sentencing, and parole decisions. The COMPAS algorithm is proprietary, so, although we know which features about defendants it considers, it is unclear exactly how it works or whether it learns over time. Still, it seems to count as AI under our broad definition. The crucial question we want to ask is whether COMPAS is fair to different groups of defendants. We will focus on Black defendants in our discussion, but the issues apply to other demographic groups as well, such as different genders.

In a now-famous analysis, the investigative journalism nonprofit ProPublica analysed predictions by COMPAS for recidivism by Black defendants in Broward County, Florida, and concluded:

> In forecasting who would reoffend, the algorithm made mistakes with Black and White defendants at roughly the same rate but in very different ways.
>
> - The formula was particularly likely to falsely flag Black defendants as future criminals, wrongly labeling them this way at almost twice the rate as White defendants.
> - White defendants were mislabeled as low risk more often than Black defendants.
>
> Could this disparity be explained by defendants' prior crimes or the type of crimes they were arrested for? No. We ran a statistical test that isolated the effect of race from criminal history and recidivism, as well as from defendants' age and gender. Black defendants were still 77 per cent more likely to be pegged as at higher risk of committing a future violent crime and 45 per cent more likely to be predicted to commit a future crime of any kind.[16]

The first bullet point says that COMPAS has a higher rate of false positives (the percentage predicted to recidivate who did not actually recidivate) for Black defendants than for White. The second bullet point then reports that COMPAS has a higher rate of false negatives (the percentage predicted not to recidivate who did actually recidivate) for White defendants than for Black. Both of these inequalities seem unfair to Black defendants.

Northpointe, the producer of COMPAS, admitted this difference in mistake rates. However, they replied by showing that COMPAS predictions are still equally accurate on average for Black and for White defendants.[17] They argued that equal accuracy yielded differences in false positives and false negatives only because the groups have different base rates of recidivism. On this basis, they concluded that COMPAS is fair to Black defendants.

To understand this complex debate, imagine that we have data about 10,000 men and 10,000 women who were all released after serving time in prison, where 8,000 of the men but only 2,000 of the women recidivated. Then imagine that ten men and ten women who are not in the above set have now been found guilty of crimes. To determine their sentence, the court asks us to predict which of these 20 are likely to recidivate in the future after prison. In the absence of any other evidence, it seems reasonable to estimate that the probability of a given man recidivating is 0.8 (or 80 per cent) and the probability of a given woman recidivating is 0.2 (or 20 per cent), because those were the rates in the past for the larger groups in the data. So we expect that about eight of the convicted men and only about two of the convicted women will recidivate in the future if not kept in prison. But now consider the rates of false positives and false negatives. All ten of the men are predicted to recidivate, because each receives a risk score of 0.8, but only eight of them actually do recidivate, so there are two false positives and no false negatives. All ten of the women are predicted not to recidivate, because each receives a risk score of 0.2, but two actually do recidivate, so there are two false negatives and no false positives. Thus, there are more false positives for the men, and more false negatives for the women, even though the predictions are equally accurate for both groups. What makes this possible is the past base rates of 80 per cent for men and 20 per cent for women in the data. If those past base rates were equal, then equal accuracy would yield equal rates of false positives and false negatives. However, the unequal base rates in past data ensure that the rates of false positives and false negatives in the present cases will not be equal when the predictions are equally accurate for both groups.

This imaginary world illustrates how it could be possible for COMPAS predictions to be well calibrated and equally accurate for both Black and White defendants (as Northpointe showed) but still yield more false positives for Black defendants and more false negatives for White (as ProPublica showed), because the base rate of crime is higher for Black people than for White people in this area. Critics of COMPAS, including ProPublica, reply that the different rates of false positives and false negatives result in significant disadvantages for Black defendants relative to White (and that these different rates arise from different base rates that are not the fault of any particular defendant who is predicted to recidivate). Both assessments of the overall fairness of COMPAS then depend on which standard of fairness they apply.

The issue at stake in these debates concerns which notion of 'fairness' is the right one to guide policy. By COMPAS's definition, AI predictions are fair when they are equally accurate for different groups. In contrast, ProPublica and other critics defined fairness by who gets the relevant benefits and burdens. By one definition, a use of AI is 'fair' only when different groups have the same rate of *bad* outcomes, such as being *denied* bail, probation, parole, or a shorter sentence. Another definition states that a use of AI is 'fair' only when different groups have equal rates of a bad outcome being *wrongly* imposed, such as bail being *denied* to those who deserve bail. A third definition requires that 'the difference between the average risk scores assigned to the relevant groups should be equal to the difference between the (expected) base rates of those groups'.[18] The list doesn't end there. Believe it or not, there are over 20 possible mathematical definitions of fairness![19]

Crucially, these definitions cannot all be achieved at the same time as long as the base rates of crimes differ between groups.[20] They all have their trade-offs, especially when also trying to reduce crime,[21] and these trade-offs might affect different racial and demographic groups to different degrees.

AIs by themselves cannot currently tell us which standard of fairness is the right one to build into policy. Nonetheless, the attempt to define 'fairness' in mathematical terms in order to build fairness into AIs can reveal conflicts among different standards of fairness and can thereby enable researchers across disciplines to become clearer about the impact of prioritizing one kind of fairness over another in various contexts. Further, if AI creators explicitly specify their definitions of fairness, it will become easier for communities to know and give feedback about whether that fairness standard is one they want to strive towards. Careful and open consideration of these issues could make it more likely that using AI for prediction will have more just consequences than if standards and goals of fairness are kept hidden and undiscussed.

Human judges versus AI: whose bias is worse?

Even if AI predictors cannot help but be unfair in some ways, it is still crucial to compare AI predictions to predictions by human judges. For example, it has been well documented that Black people are disadvantaged by the current judicial system even when AI is not used. Human judges have their own implicit and explicit biases about race (and other demographic characteristics of the defendants),[22] which can affect their decisions even when they are truly doing their best to be impartial.[23] So is AI better than human judges? The discussion comments on the Pro-Publica article framed the issues this way:

> [*Commenter B*]: What is scary is that the results of this program [using COMPAS in Broward County] have been shown to be inaccurate and racially biased (even after controlling for different rates of crimes between certain races).
>
> [*Commenter K*]: Even scarier is when 10,000 judges across the country make decisions where no one can see their 'algorithm' and bias – and we just let them continue to perpetuate injustice. I prefer an algorithm that everyone can see, study, and work to fix. It's easier to fix and test the algorithm than to train and hope judges don't bring bias into decision-making.[24]

At this point there really isn't enough evidence to make definitive conclusions about when human judges or AI systems are more biased, and this comparison might well change with context and as AI develops.

Further, even if an AI is less biased, human judges can still be biased in how they apply or reject the AI's recommendations. There was evidence of this in Kentucky, where the introduction of algorithmic predictions of bail violations was unexpectedly correlated with judges offering no-bail release to White defendants more often than to Black defendants because the judges were more likely to overrule the algorithm's predictions for moderate-risk defendants if the defendants were Black.[25] Nonetheless, as 'Commenter K' suggested, the use of AI might offer some benefits in the pursuit of racial distributive fairness, if humans use it properly.

POTENTIAL FOR TRANSPARENCY

Because COMPAS's system is proprietary, judges, lawyers, and defendants cannot find out how features of the defendant are combined to predict recidivism. Moreover, as has been mentioned elsewhere, many AI systems function as 'black boxes' whose reasons for making predictions are very difficult, if not impossible, to discern. For such reasons, AI predictions are sometimes opaque.

Nonetheless, other AI systems are explainable and interpretable, while still providing good prediction performance.[26] When interpretable AI is used for judicial decisions, it can be made clear which factors influence its decisions and how they influence it, even if making this clear requires some translation by AI experts. This in turn allows those factors to be scrutinized, challenged, and addressed. As Commenter

K suggested, this relative transparency seems like an advantage over judicial decisions which are strongly influenced by human biases that are never officially acknowledged. We will address this in more detail shortly.

EXPLICIT PREJUDICE AND INDIRECT PROXIES

Another possible advantage of using AI to predict recidivism is that, whereas judges will usually know a defendant's race, AIs can be intentionally designed to avoid using racial or other demographic categories in its predictions. When designed this way, AIs don't use racial categories directly as predictors of recidivism, allowing them to avoid some patterns of explicit racial prejudice that some human judges could have, even though most judges sincerely aim to be impartial.

Unfortunately, however, even if an AI is not given racial information directly, the data that it analyses can still include information about other categories that are highly correlated with racial categories (called 'proxies'). For example, almost all recidivism prediction algorithms are trained on data about arrests instead of crimes. Police sometimes monitor certain racial communities more closely, which leads to them making disproportionately more arrests in those communities, even if many of those who are arrested are eventually released without being charged with a crime. As a result, some AI systems (such as COMPAS) that do not use race explicitly, but are trained on arrest rates, can end up assigning unduly high risks of recidivism to defendants who are members of underprivileged racial groups. This problem is an example of the infamous 'bias in, bias out' issue we discussed at the beginning of the chapter. Bias in our societal structures and police procedures will lead to bias in data used to train AIs ('bias in'), which will in turn lead AI algorithms to predict too much risk of recidivism for members of certain disadvantaged communities ('bias out').

This kind of indirect discrimination can solidify unjust social structures or hierarchies and make them harder to change. As former US attorney general Eric Holder has said, 'By basing sentencing decisions on static factors and immutable characteristics – like the defendant's education level, socioeconomic background, or neighborhood – they may exacerbate unwarranted and unjust disparities that are already far too common in our criminal justice system and in our society.'[27] It definitely seems unfair to sentence defendants in ways that make inequities worse. There is still room for hope that a combination of technical advances and careful design of the systems in question can help AI combat this type of pernicious bias, though. We will discuss these possibilities next.

CORRECTIONS AND PROTECTED CLASSES

In theory, AI algorithms should be able to leverage their quantitative models of the world to statistically correct for certain unfair outcomes, at least to some degree. Different technical strategies are being developed to do this.[28] To give an idea of how they might work, we will describe one of the more straightforward approaches. It has three steps. First, a large data set is used to train a statistical model to try to predict something we care about (like bail violations or recidivism) using all of the information we think might be relevant, including racial and other demographic categories. This model will have learned 'weights' for each feature, including 'protected' features like race. Second, the impact of the protected categories is statistically removed from the other features used to make predictions in the model. In the case of race, the result is 'race-blind' weights, though the weights can be made 'blind' to any demographic category in the data. Third, predictions about specific cases are made using the 'race-blind' weights instead of the original weights, and the weights from racial features are excluded altogether.[29] The result is a prediction 'corrected' for racial bias where the statistical influence of race on the prediction has been removed (to a large extent).

Supporters of risk algorithms 'corrected' in this way claim that would make it more likely that Black and White individuals who are otherwise identical will receive the same risk score or prediction. This method is supposed to reduce the adverse differential impacts of the algorithms on disadvantaged minorities. In illustration, a simulation demonstrated that using this bias correction method on New York City bail decisions would have resulted in 1,700 more Black defendants being released on bail during the analysed time period, leading to only eight additional failures to appear at trial, compared to decisions that were made without any bias correction.[30]

Some critics still object to any prediction method, like this one, that explicitly uses protected labels, like race, because they see this reliance on racial classifications as a form of discrimination. In response, others argue that it is not only acceptable but praiseworthy to use protected categories with the goal and effect of reducing unfair disparities between the labelled groups. In addition, methods like the one we described use protected categories only in the model training or pre-processing steps, so they are not used in the final step of making predictions about a specific individual. On this basis, many legal scholars have argued such approaches would not violate any civil rights and might even help to protect them.[31]

What about procedural justice?

So far, we have focused on how AI can affect whether certain benefits and burdens are distributed fairly to different groups. This kind of fairness is distributive justice, the first of the three kinds of justice we described at the beginning of the chapter. It can also be important to consider the second kind of justice – retributive justice – if its predictions lead to people receiving sentences that are too harsh or too lenient for their crimes. The third kind of justice mentioned earlier is also important. Even if a defendant is correctly found guilty, receives an appropriate sentence, and punishments are distributed fairly between demographic groups, a defendant can still be treated unfairly in a different way if the trial *procedures* are not fair. So now we need to ask: even if AI optimists win out and the legal system ends up using AIs that are shown to be sufficiently fair distributively and retributively, could those same AIs still be procedurally unjust or unfair?

Among other things (such as an impartial judge and a speedy trial),* procedural justice in law is usually thought to require a right for each side to cross-examine the other's witnesses and, more generally, to question their evidence. Each side must be able to understand the other's witnesses

---

* Another aspect of procedural justice in many systems of criminal law is a right to a jury of our peers. Is AI our peer? We will not discuss this issue here, but it might be coming soon to a court near you.

and evidence for any cross-examination to be effective. This ability to question becomes a critical issue when AI predictions are a basis for legal decisions. If the AIs that made those predictions are unintelligible to anyone other than an AI expert, or if they are impossible even for experts to understand, then the defence loses its ability to respond effectively. That would make court procedures unfair.

This argument was made about an AI-informed sentencing decision (rather than a bail decision) in *Loomis v. Wisconsin*.[32] Eric Loomis was charged with taking part in a drive-by shooting. He denied firing the shots but pleaded guilty to 'attempting to flee a traffic officer and operating a motor vehicle without the owner's consent'. Before a COMPAS score was introduced into Loomis's case, the prosecution and defence had agreed upon a plea deal of one year in county jail with probation. At sentencing, though, a probation officer shared that the COMPAS AI predicted Loomis would probably re-offend. The trial judge stated, 'You're identified, through the COMPAS assessment, as an individual who is at high risk to the community. In terms of weighing the various factors, I'm ruling out probation because of the seriousness of the crime and because your history, your history on supervision, and the risk assessment tools that have been utilized, suggest that you're extremely high risk to re-offend.'[33] Loomis was then sentenced to six years in prison and five years of extended supervision.

Loomis appealed the sentencing decision. One of his critical arguments was that his trial was unfair not only because COMPAS was unfair to certain groups, but also because COMPAS's predictive model was both proprietary and complicated (being based on 137 questions), so there was no realistic way for Loomis or his attorney to know how or why COMPAS arrived at its risk prediction or to cross-examine, understand, or respond to its prediction. Loomis ultimately lost his appeal, but many legal scholars think he should have won, particularly because of this procedural argument.

The procedural right to know why and how COMPAS is labelling people as 'likely to reoffend' is important not only to defendants. COMPAS's inner workings are important for judges as well.[34] The trial judge in Loomis's case needed to be able to make informed decisions about when (and how much) to trust COMPAS's predictions in order to be justified in believing that Loomis was truly 'extremely high risk to re-offend'. Without this knowledge, the judge would need to accept or reject the algorithm's prediction blindly and could end up confidently following the prediction even when it is unreliable. Appeals court judges would be in the same position and would have no way to know whether trial court judges made a mistake by following COMPAS's predictions, or how to correct those mistakes accurately. Legislators also need to know enough about what COMPAS is doing under the hood and how it performs in different groups to be able to discern whether there are problems with how the judicial system is using it that require new legislation. Even regular citizens need such knowledge in order to choose when to advocate for legal and political reforms. In all of these ways, judicial uses of proprietary, unexplainable, opaque, or black-box AI without vetted accuracy records can contribute to unjust and unfair procedures in our legal system, even if AI algorithms that are used distribute punishments and benefits fairly.

CHAPTER 4

Does interpretability solve the problem?

Algorithms are considered interpretable when humans can figure out what caused them to produce their outputs. If we required all AIs used in the justice system to be interpretable, and also required the developers of such AIs to share how their AIs were trained and how they work, would that remove all concerns about the procedural justice of these AIs?

It would help a lot, but there will still be some challenges. Black-box deep learning AIs are popular because they often perform better than any other currently known AI technique. Interpretable algorithms are sometimes less accurate than uninterpretable algorithms, and this inaccuracy really matters when it comes to decisions that can affect whether somebody will be put in jail and for how long. The good news is that this technical area of research has both a successful track record and a lot of room for growth. Some prominent computer scientists have argued that properly constructed 'interpretable' algorithms can do as well or nearly as well as black-box algorithms in almost all of the cases considered here, and a considerable amount of research supports this view.[35] If more energy and money is invested in such research, it is realistic to hope that many interpretable AIs can be made as accurate – or almost as accurate – as any black box.

Another complication is that 'interpretability' means different things to different people.[36] Even if a computer scientist can understand and predict how an interpretable algorithm will behave, that doesn't mean a typical lawyer or defendant will be able to understand it or predict its behaviour. What kind and what degree of intelligibility is required for a fair legal system? No one answer will fit all people or all roles in the system. AI developers will need deep understanding of how an AI is making its predictions and the details of the data that were used to train the system. They need this so that they can make or call for any modifications required to ensure the AI is fair and sufficiently accurate. In contrast, judges and legislators need to know only a little bit about how the AI works and was trained, but they also need to know what the purpose of the AI is for the legal system so that they can assess how well that purpose is being served and then request changes when necessary. Thus, whichever AI algorithm is chosen, it needs to be made intelligible to these different players in different ways.

It is very difficult to figure out how to present and describe information about AI models in ways that stakeholders can make sense of. Further, we do not always use AI models in our decision-making in ways we would expect. For example, recall that AI recommendations were linked with judges in Kentucky making *more* biased bail decisions because they rejected the AI's recommendations in a biased way.[37] In addition, providing explanations of how an AI algorithm works doesn't necessarily make people any more accurate at identifying the AI's faulty predictions.[38] So there's a long way to go before we figure out how to make an 'interpretable' AI's decisions sufficiently accessible to everyone who needs to evaluate them. Even as we make progress, some types of procedural unfairness are likely to remain. Nonetheless, requiring AI algorithms to be as interpretable and transparent as possible will go a long way towards enabling them to be more fair procedurally and in other ways.

Fair AI

Above, we focused on AI unfairness across racial categories in criminal law to explore the different ways in which AI can threaten, or strengthen, justice and fairness. Of course, AI can have similar impacts on countless other areas inside and outside of law. Whenever AI is used by civil courts to assess future damages; by banks to predict whether a loan applicant will default; by employers to hire, fire, or promote workers, or to award bonuses; by schools to screen applicants; or by websites to predict what products to recommend to you – these fairness concerns will be relevant and there is a risk of harm, including subordination and stigmatization, to disadvantaged groups. Further, discrimination is obviously directed at a variety of socially salient groups, including those defined by ethnicity, gender identity, and sexuality; the poor and homeless; and the elderly and disabled. AI needs to be fair across many demographics at once. That will not be easy, and AI creators will need to be perpetually vigilant about AI fairness issues and constantly monitor their AI systems for potential bias.

The silver lining in all of this is that the introduction of AI across so many aspects of life has helped to make more of us aware of many forms of injustice in the decisions that humans have traditionally made. Even if we haven't yet figured out how to apply AI fairly in all circumstances, at least AI is highlighting unfairness that needs to be addressed.

Moreover, AI fairness might become easier to manage.

Progress has already been made in generating technical 'fair AI' tools that can help address and mitigate AI unfairness issues in scalable ways. The annual conference on AI 'fairness, accountability, and transparency' illustrates how much commitment and momentum there is in this arena. To give you a flavour of what is being developed, in addition to some of the technical correction techniques we already mentioned, tools have been made to audit the representation of disadvantaged groups in data sets,[39] and multiple entities are generating 'fair' data sets for other entities to use which are confirmed as having acceptable representation of diverse demographic groups.[40] Auditing tools and services have also been developed to monitor the outcomes of AI models on groups after the models have been deployed, so that unfair impacts can be identified and addressed efficiently.[41] In addition, many frameworks, checklists, and organizational approaches are available to help AI teams identify unanticipated fairness issues in their AI products.[42] Some AIs are even being proposed to help AI teams identify which mathematical definitions of fairness are most appropriate for their use case.[43] Again, these are just a small subset of the 'fair AI' technical approaches being developed, and organizations and governments are increasingly adopting these tools.[44]

Technical tools are still not sufficient to address all of the ways in which AI can be unfair. Many of these tools are hard to use, or it's hard to figure out how to apply them to specific cases. There are still significant gaps in what is provided by the tools and what is needed for people to implement them successfully.[45] More fundamentally, many societal and organizational challenges also have to be addressed in order for AI technical tools to be implemented correctly, as we will discuss in Chapter 7.[46] Nonetheless, the potential for AI to improve on human biases suggests that properly implemented AI might benefit society, despite its significant but manageable risks to fairness.

CHAPTER 5

Can AI (or its creators or users) be responsible?

In the dark of night, around 9:58 pm on Sunday, 18 March 2018, 49-year-old Elaine Herzberg wheeled her bicycle across the poorly lit Mill Avenue in Tempe, Arizona. Meanwhile, 44-year-old Rafaela Vasquez was streaming an episode of the television talent show *The Voice* while working as a test driver in the driver's seat of a self-driving Volvo XC90 owned by Uber, travelling at 43 miles per hour on the same road. The sensors and AI detection system in the car failed to identify Elaine Herzberg as a human pedestrian, so the car's brakes were not engaged sufficiently or soon enough to avoid hitting her. Vasquez could have slammed on the brakes herself but was looking away from the road for a few seconds before the car crashed into Herzberg and killed her.[1]

Who or what was responsible for Herzberg's death? Herzberg? Vasquez? The car? Uber? Safety managers at Uber? Engineers at Uber? Arizona government officials who allowed Uber to test their cars in Tempe? All of them? None of them?

Here's why we should we care. Herzberg's death is a paradigm example of the type of AI safety issue we discussed in Chapter 2 which can have negative ripple effects on society if not curtailed. However, if nobody is held responsible or punished for it, there will be little incentive for anybody contributing to AI technology to prevent similar accidents from happening in the future.

To prevent these undesirable upshots, we need to figure out who is responsible for accidents and harms involving AI. The problem is that it is not at all clear how to do that. To give a sense of the challenge, let's think through who or what should be held responsible for the accident that caused Elaine Herzberg's death.

What is responsibility?

Imagine that, during a famine, two parents leave their children and walk to another village to get food for themselves. Meanwhile, their children die from starvation. Someone might say, 'The drought was *responsible* for the famine. Still, the parents had a *responsibility* to feed their children. The courts will not hold them *legally* responsible. Nonetheless, they were *morally* responsible for their children's deaths.' The first sentence means simply that the drought *caused* the famine. The second sentence means that the parents had a *duty* or *obligation* to feed their children (presumably because of their role as parents). The third sentence means that the courts will not impose civil *damages* or criminal *punishment* (though perhaps they could or should). The fourth sentence means that the parents deserve or are liable to negative moral *sanctions*, such as blame and anger. These are all different notions of responsibility.

The second notion – responsibilities as duties and obligations – is important to any well-meaning AI researcher or developer who wants to act morally even when they know they can get away with something. Our final chapter will discuss some ways to train people and structure institutions so that they are more likely to fulfil their responsibilities in this second sense.

This chapter will instead focus on the third and fourth types of responsibility – liability to legal punishments or moral sanctions – because for AI to have a net positive impact on society there need to be reliable mechanisms to minimize its harms, and sanctions are the primary way we can incentivize entities to avoid those harms. So, for our current discussion, we will worry about the first two kinds of responsibility – causes and duties – only insofar as they affect liability to punishments and sanctions.

What is the difference between legal and moral responsibility? Legal responsibility is primarily related to whether an entity is liable to punishments, such as fines or imprisonment, prescribed by authorities with special legal powers. Moral responsibility, in contrast, can be applied by people without any special position and according to rules that are not created by any special authority. Moral sanctions come in forms like public blaming, social or corporate ostracization, and moral outrage (by victims and observers). Some philosophers, such as John Stuart Mill, also include an agent's own shame, guilt feelings, and felt obligation to apologize as 'internal' moral sanctions. Often, legal systems punish people for things that they are morally responsible for. However, some legal systems impose legal sanctions on people who are known not to be morally responsible (for example, when a vendor sells a defective item that they genuinely had every reason to think worked fine). It is also possible to hold people morally responsible for things that are legal (such as lying to a friend). That's why it is important to distinguish between law and morality.

Putting this all together, when we ask who is responsible for harm involving AI, we are really asking who or what is liable for sanctions, where the sanctions could be legal or moral. Importantly, many entities can be responsible at the same time. Further, different entities can be responsible for different reasons, in different ways, and to different degrees. If they are only partly responsible, this can make them liable to some sanctions but not others, such as when they deserve public criticism but not legal punishment, or they deserve private criticism but not public rebuke. Partial responsibility can also imply that an agent is liable to pay compensation for some but not all of the damages that were caused. With these distinctions in mind, we are ready to tackle the task of deciding who is responsible for Herzberg's death on 18 March 2018.

Was the human driver responsible?

Some people believe that Rafaela Vasquez was responsible for Elaine Herzberg's death because she was in the driver's seat of the autonomous car, was working as a test driver, had been instructed to pay attention and take control of the car when needed, and could have hit the car's brakes earlier if she hadn't been looking away from the road when Herzberg was crossing the street. Vasquez knew (or should have known) that there was some danger of a person or object appearing in front of the car, but she did not sufficiently heed that risk. As a result, she was not looking at the road for several seconds before the crash and did not see

Herzberg or brake in time to save her life. This all makes Vasquez seem responsible.

What defence could Vasquez present? The accident happened very quickly, so Vasquez might argue that she could not have prevented it even if she had been paying full attention to the road. This claim is somewhat supported by the video, since Herzberg becomes visible in the headlights only a second or two before the crash. It is not clear whether the collision was completely unavoidable, though. At least one police analysis argued that Vasquez could have stopped more than 42 feet (13 metres) in front of Herzberg if she had been looking at the road and had responded appropriately.[2] Still, it is not clear whether a driver of a regular car would be held responsible in a similar situation if they looked away from the road for a few seconds to adjust the radio.

Another defence posed by Vasquez's lawyers is that the company required her to monitor its Slack channel* and diagnostics about the car,[3] and that was why she wasn't looking at the road during the seconds before the accident. According to her attorneys, she was listening to *The Voice* at the time, but she was not watching it. Instead, it was only background noise, like you might hear from the radio. They claim the dashboard video makes it clear that *The Voice* was being played on her personal phone in the passenger seat, and Vasquez was looking not at that phone but rather at the work-related screen in the centre console just before the accident. Uber confirmed that its drivers were supposed to

---

* Slack is a digital instant messaging platform, and Slack channels are dedicated chat spaces for specific groups of people.

check work Slack channels on a regular basis, but claimed that drivers were instructed to do so only on their breaks or when they pulled over. Vasquez, in contrast, argued that drivers were expected to monitor Slack channels in real time. In essence, Vasquez claims that the horrible accident was a result of her doing what her employers told her to do.[4]

Vasquez also claims that her employers misled her about the car's brakes. The type of car Uber used as the basis for its self-driving car had an auto-braking feature created by Volvo which would engage the brakes automatically if a collision was likely. This system had been disabled by Uber engineers because it supposedly interfered with Uber's AI software, but Vasquez's lawyers claim that her employers told her that it was still functional as a back-up safety system in the automated car she was driving. It is not clear whether anyone at Uber really said that. But if Vasquez truly was led to believe that the Volvo anti-collision system was active in her car, perhaps she was justified in trusting the car enough to take her eyes away from the road for a few seconds, especially if she thought she needed to stay on top of company Slack messages to keep her job.

Vasquez might have additional defences, too, given the unique nature of AI-driven cars. Even if Vasquez could have seen Herzberg coming, Vasquez could claim that she was not really a driver, because a self-driving car drives itself. In other words, Vasquez might have been more like a passenger in a taxi than a taxi driver. If a taxi driver runs over a pedestrian, the passenger is usually not responsible. By analogy, perhaps Vasquez is also not responsible when the AI system runs over a pedestrian. The obvious response to this argument is that Vasquez is not merely a passenger, because she was able to take over control of the car and was both paid and instructed to take over the car in dangerous situations. Still, we should consider whether the inclination to hold Vasquez responsible exists largely because we don't know whom else to blame.

We also need to consider whether it is realistic to expect human drivers to be sufficiently vigilant to intervene in every case where a self-driving car makes a mistake, or to follow instructions to watch the road at all times. When things rarely go wrong, it is almost impossible for humans to stay fully engaged with a monitoring task. For this reason, Waymo, one of Uber's competitors in the self-driving car race, decided that it was too dangerous to rely on humans to intervene with self-driving cars at all, so they made the strategic decision to make their AI-driven cars fully autonomous.[5] Perhaps Uber had the responsibility to make the same decision, and it is not fair to blame Vasquez when many of us would have done no better.

We are all very susceptible to automation complacency: as soon as any part of a task is automated, human overseers trust the automated system to get the job done and therefore stop paying attention. Given that Vasquez had taken the self-driving car to that same location in the past without incident, it might be understandable that Vasquez trusted the car to be safe enough to allow her to take her eyes off the road for a few seconds. If most people would do the same in her situation, perhaps Vasquez is just unlucky instead of responsible.

Vasquez could clearly offer many arguments in her defence, but a lot hinges on what she was told, how she was trained, and what was in the contract she signed. If she was told the system was very reliable, so she needed to monitor it only once in a while, then she would seem less responsible than if she was told to keep her eyes glued to the road at every moment. The same goes for the question of whether Uber told her to monitor company chats while driving or had a culture that rewarded monitoring company chats while driving. Without this information, it is hard to reach any definitive conclusion about the depth or breadth of her responsibility.

Was the pedestrian responsible?

Even if Vasquez was responsible to some extent, Elaine Herzberg might also be responsible for acting in an unsafe manner as a pedestrian before she was hit. Herzberg was wearing dark clothing and had no reflectors on when she crossed the road. Her bike did have front and rear reflectors and a forward headlamp, but these were not visible to vehicles coming perpendicular to her path as she crossed the road. Herzberg chose to illegally cross where there was no designated pedestrian crosswalk, even though there was one about 360 feet (110 metres) away and four signs warning pedestrians to cross only at the crosswalk. In addition to all these risks, the video from the Uber car shows Herzberg walking her bicycle slowly across the road without looking to the side for cars until just before impact. Herzberg and her daughter also reportedly crossed at the location of the accident frequently, because they often camped in the area and crossed the street to charge their phones at an electrical plug nearby. Further, Herzberg is reported to have had methamphetamine in her blood when she died, which might have led her to make different choices than she would have made had she not been taking methamphetamine.[6]

Putting all this together, Herzberg made many risky choices when crossing the road with her bike, so it seems like she is at least partially responsible for taking the risks she did. Nonetheless, it also seems clear that Herzberg does not deserve all the blame for what happened, especially because part of the lofty promise of self-driving cars is that they should be able to prevent exactly these types of accidents.

Were Uber's AI contributors responsible?

Maybe the people who should be liable are those who turned the Volvo car into a self-driving car by, among other things, picking and installing sensors, and creating an AI that interpreted their signals and controlled the car accordingly. We will refer to this group as Uber's AI contributors. Remember that AI systems are made by many different types of both technical and non-technical contributors, and each contributor understands each part of the system to varying degrees. Some parts of the AI system may have been bought from a third party, perhaps leading to Uber's team not knowing all the details of how those purchased components worked. If specific AI contributors *did* know that the sensors or AI in the car Vasquez was driving could malfunction in unacceptably dangerous ways, though, it could be argued that those contributors should have tried to prevent Uber's self-driving cars from being put on public roads until the system performed in an acceptable way. In legal language, these contributors could be deemed to be responsible through 'recklessness' because their continued intentional contribution to putting Uber's self-driving cars on the road, despite knowing the risks, reflected insufficient regard for the safety of others, the same way drunk driving reflects insufficient regard for the safety of other people on the road.

One employee did email Uber executives days before the accident saying: 'The cars are routinely in accidents resulting in damage. This is usually the result of poor behavior of the operator or the AV technology. A car was damaged nearly every other day in February. We shouldn't be hitting things every 15,000 miles.'[7] This documented warning does strongly suggest that many of Uber's self-driving car contributors must have known that the cars could behave in unsafe ways and/or frequently got into accidents, and they not only didn't try to stop the cars from being put on the road, but also actively contributed to processes that would put the cars on the road in their known unsafe state. Contributors who had sufficient knowledge of the cars' risks *and* sufficient power to get the cars to be used more safely are at least partly responsible for Herzberg's accident, even if it's difficult to figure out which contributors met those criteria and impractical to charge them legally.

Another way Uber's AI contributors could be responsible is through negligence, or a failure to behave with the level of care and attention that would be expected of someone in the same role under the same circumstances. Whereas recklessness generally requires a wanton disregard for others' safety, negligence is a lower standard, requiring only that offenders commit unacceptable commissions or omissions, perhaps leading to them to be insufficiently aware of risks they would otherwise try to prevent if they knew about them.

Many circumstances surrounding Uber's self-driving cars suggest that the company's engineers and contributors (along with its executives, lawyers, and others) might be responsible through negligence. For one thing, automobile accidents can end lives, so people who design or deploy such life-threatening systems are usually expected to exercise high levels of care and attention and to anticipate all kinds of risks, even ones that would occur only rarely. Nonetheless, according to an investigation by the National Transportation Safety Board (and confirmed by Uber itself), Uber's system was not designed to consider the possibility that pedestrians might jaywalk, i.e., walk across roads outside designated crossings, or push bicycles across the road while on foot. These oversights were why the car never identified Elaine Herzberg as a human pedestrian walking a bicycle across the road and, instead, got hung up trying to decide what kind of object was ahead of it instead of promptly slowing down once an object was detected.[8]

People jaywalk and walk their bikes across roads all the time. It is bizarre that engineers of a system designed to navigate normal streets, especially streets in urban locations, would fail to plan for these standard human behaviours. Even minimally competent professionals in the auto-driving industry should be able to anticipate such scenarios on their own. If not, though, one would expect that any reasonable review of the cars' performance would identify jaywalking and walking a bike as scenarios the cars need to be able to navigate safely. According to the National Transportation Safety Board report, Uber test drivers had even reported jaywalkers on the exact same route Vasquez was driving the day of the accident. This makes it hard to imagine a good justification for Uber's failure to prepare and evaluate its automated systems for the kinds of pedestrian behaviour Herzberg exhibited.

Further, Uber's self-driving system was intentionally designed so that it would not press the car's brakes fully unless it could label an object in front of it and was sure the obstacle could be fully avoided. If the car decided it would use hard braking, it would wait one second before actually engaging the brakes, reportedly to allow the system time to verify the emergency, minimize the inconvenience of false alarms, and allow a human to take over. We couldn't find any documentation that explained how Uber would use the one-second delay to achieve these goals. In contrast, if the car determined that an obstacle in front of it could *not* be fully avoided, rather than engage hard braking to stop the car as much as possible before impact, the car would only slow down gradually and warn the operator.[9]

If you are confused about why a self-driving system would be designed this way, you are not alone. These design choices are at best odd and very possibly completely careless. One can imagine scenarios that might motivate the car designers to put limits on when brakes are engaged. For example, if human drivers are so alarmed by automatic braking that they overcorrect the car's steering in a dangerous way, perhaps it is safer to engage the brakes more gently at first so that the human drivers can intervene with less anxiety. However, it's hard to imagine a reasonable justification for designing a self-driving car that requires human intervention to stop it when it is about to collide with something. Even if Uber's

AI contributors are not legally liable for these choices, it does seem likely that they were morally responsible for their neglect of these safety issues in the Herzberg accident.

Another issue is that, as we have seen in other scenarios in previous chapters, some AI systems use algorithms that lead to black-box models that are neither explainable nor interpretable to the AI developers – let alone to other contributors. Recall from Chapter 1 that many of the best-performing AIs, especially deep learning AIs, use these types of models. It is not clear whether Uber's AI contributors used black-box models that are neither explainable nor interpretable. If they did, they might not have been able to anticipate when the model would make mistakes, nor been able to understand why the self-driving cars made certain mistakes, including not being able to classify a person walking a bike across the road outside a crosswalk. Choosing to use only explainable or interpretable AI could have mitigated against such problems.

However, the opacity of AI models matters in relation to responsibility for specific harms only if someone involved needs to know the details of how a model works in order to predict that those harms will occur. It is not clear that this is relevant in the Uber case because, according to the email sent to executives and subsequent whistleblower accounts, many of Uber's AI contributors knew the cars were routinely in accidents resulting in damage. Given this knowledge, Uber's self-driving car team should have anticipated and routinely monitored for the mistakes the Uber car made. Still, the opacity of the AI systems could influence the AI contributors' responsibility for some aspects of what led to Herzberg's death or their responsibility for other instances of self-driving car accidents.

Regardless of whether you agree that Uber's AI contributors were morally responsible for Herzberg's death in some way, the contributors are unlikely ever to be charged, because Uber settled out of court.[10] In general, individual technology contributors are usually not held legally responsible for technology harms or failures unless they break their employment contracts, violate their company's ethical guidelines, or avoid the steps a company put in place to avoid the harms. Instead, the company that employs the AI engineers and AI technology contributors is typically held legally liable. This liability approach helps circumvent the difficult process of figuring out exactly which role a given contributor played, allows individual technology contributors to innovate more freely without needing to be prohibitively concerned about their legal liability, and provides deeper pockets to compensate victims of any harm that occurs.

Was Uber responsible?

For the reasons mentioned above, a company or employer is typically legally responsible for the actions of its employees, as long as the employee's employment contract doesn't state otherwise and the employees act within the course and scope of their role at the company. This is generally true even if a company's leaders do not know the technical details of what their employees are working on. Since Rafaela Vasquez was working for Uber when she was involved in the accident, Uber could potentially have been held legally responsible for what happened.

There are many reasons to think that Uber was responsible for Herzberg's death. Even if the company's leadership didn't intend their self-driving cars to cause harm and could not have predicted the exact time or location of the accident, Uber still might have been either negligent or reckless in a way that contributed to the accident. Then the company could be both legally and morally responsible.

After all, Uber's decision-makers must have been aware of the risks created by their actions, given the many prominent societal conversations about such risks. If Uber's decision-makers were unaware of the risks, it could only be because they recklessly disregarded warnings from industry experts. Recall that there was also at least one documented email from an employee to Uber executives warning of the cars' frequent accidents. Thus, Uber's executives had to have known about the risks of putting their self-driving cars on the road.

Despite knowing about these risks, Uber disabled Volvo's auto-braking system. According to subsequent studies, the Volvo-installed system would have stopped the car before it hit Herzberg, had it been functional.[11]

Why did they take these risks? Perhaps Uber executives' enthusiasm for getting the cars on the road as quickly as possible was motivated by a belief that making vehicles autonomous would ultimately save human lives. Alternatively, maybe Uber decision-makers were motivated by the promise of financial profit. Either motivation might be justifiable as long as sufficient efforts were made to protect human lives while developing the technology. Of course, there can be valid debate about what level of effort is sufficient, but the National Transportation Safety Board reached consensus that Uber's safety culture and practices reflected disregard for human life. At the time of the accident, Uber did not have a safety division, a safety plan, a dedicated safety manager, or people in charge with experience of managing safety. These are glaring omissions for a company working in a safety-critical domain. Uber also did not have a fatigue-management programme, which is a standard component of most transportation companies, and it failed to adequately monitor its drivers' performance during training or test driving. Despite Uber having installed cameras to track what drivers were doing, managers rarely examined that footage or provided feedback to those drivers about unsafe actions. In addition, Uber's management made decisions that seemed to prioritize savings over safety, such as reducing the number of human drivers in their cars from two to one.[12]

Some have accused Uber of intentionally designing its systems and practices so that its drivers would be found legally at fault for accidents instead of Uber.[13] In the words of a whistleblower, Uber was 'very clever about liability as opposed to being smart about responsibility'.[14] Even if efforts to avoid liability for safety failures were not so nefarious, the conclusion of the chairman of the National Transportation Safety Board was that 'Safety starts at the top . . . The collision was the last link of a long chain of actions and decisions made by an organization that unfortunately did not make safety the top priority.'[15] Therefore, Uber seems to have some moral responsibility for Herzberg's death.

Is Uber the only company responsible? Volvo allowed Uber to use its car and disarm its anti-collision system. Nvidia supplies chips for autonomous cars. If these companies knew Uber was using their products in ways that could ultimately lead to unjustified harm, they might be partly responsible for Herzberg's death as well. Nvidia had its own self-driving cars on the road in 2018, in addition to providing Uber with chips. It chose to temporarily suspend its own self-driving tests after the accident, stating, 'Although we developed our self-driving technology independently, as good engineering practice, we will wait to learn from Uber's incident. The entire industry will learn from this incident.'[16] Thus, Nvidia recognizes a professional obligation for technical AI contributors as well as others in the company to take safety data into account when they decide how their AI is used.

Still, no company could simply test drive its cars in Arizona on its own, at least not legally. It had to be given permission to do so by Arizona officials. What about their responsibility?

Was Arizona's government responsible?

Governments are supposed to safeguard the wellbeing of their constituents, taking into account physical as well as financial and emotional wellbeing. They can provide incentives to support things that help their people, and they can also take action to regulate things that could harm them. Since at least some Arizona government officials should have known that self-driving cars pose accident risks and had the power to prevent Uber's vehicles from being driven on the state's roads, are they at least partly responsible for any harm caused by crashes Uber's self-driving cars get into?

In the case of Vasquez's crash, the situation was complicated. Arizona's governor at the time, Doug Ducey, was a champion for self-driving car technology, and openly wanted to cultivate more autonomous vehicle activity and research in his state. He believed self-driving cars would save lives in the long run, and wanted to capitalize on opportunities for the industry to create jobs and income for Arizona residents. This is all very reasonable, and many would applaud him. Still, some of his decisions as governor were controversial. The most famous pertains to yet another wrinkle in the Uber crash story.

In 2016, Uber refused to apply for special permits to test autonomous vehicles on California roads, despite announcing with great fanfare that it would begin testing its self-driving cars in San Francisco. Uber claimed it didn't need to apply for the permits because it required a driver to be present in its cars at all times, meaning they were not autonomous under the definition of California's Department of Motor Vehicles. Instead, it was argued, Uber's self-driving cars should be treated just like any other car. The company's reason for this resistance was strategic: if Uber's self-driving cars were not classified as autonomous, Uber would not be required by California law to track and report the vehicles' accidents. Other autonomous car companies that required human drivers for safety reasons applied for autonomous vehicle permits without protest, and received them easily. Uber nonetheless stood its ground, continued to avidly promote its cars as 'self-driving', and proceeded with testing the cars in downtown San Francisco without the required permit. Within hours, the cars were documented running red lights and getting into many other unsafe situations, and these safety mistakes were widely publicized.[17] Uber continued to refuse to apply for an autonomous vehicle permit, which led to California revoking the registration for Uber's self-driving cars and threatening legal action against Uber if its self-driving cars were not taken off the road. As a result, Uber stopped piloting its cars in California.

Almost immediately after all this went down, Ducey encouraged Uber to test drive the exact same cars with the same safety records in Arizona instead. He famously said: 'Arizona welcomes Uber self-driving cars with open arms and wide open roads.'[18] In contrast to California's rules, which required all permitted autonomous vehicle companies to report crashes and the number of times human operators had to intervene when test driving their cars on public roads,[19] Arizona's rules did not require any safety procedures or reporting from autonomous vehicle companies. Arizona did create a self-driving vehicle oversight committee of experts to advise state agencies, but that committee met only once, and it never made any recommendations or took actions to evaluate autonomous vehicles' safety record rigorously (such as calling witnesses or demanding documents). In addition, some have alleged that Ducey's office encouraged Uber to test drive in Arizona while intentionally keeping Arizona citizens in the dark about the presence of autonomous vehicles on the state's roads, because it wanted to circumvent protests from concerned citizens, especially given the safety mistakes Uber's cars had made in San Francisco.[20] Furthermore, it has been suggested that Ducey's office may have had a 'cosy' professional and maybe financial relationship with Uber which impaired its ability to make decisions that sufficiently considered citizens' interests, and which incentivized officials to treat Arizonans like 'lab rats'.[21]

So, like all the entities we have discussed so far, the Arizona government didn't intend Herzberg's death and couldn't have predicted this particular accident. Nonetheless, they knew or should have known that Uber's self-driving cars posed many risks, but they allowed Uber to create those risks. If Ducey's treatment of Uber's safety risks was either negligent or reckless, he would have some responsibility for Herzberg's death.[22] The same goes for anyone in the Arizona government who knew about the risks of Uber's technology and knew that Uber's cars were going to be tested on Arizona's roads but did nothing to stop the testing or make it safer. Whether the Arizona government's role in Herzberg's death is best explained by such negligence or recklessness is central in the lawsuit raised by Herzberg's family against Arizona (as well as against the local Tempe administration).[23]

Is the AI responsible?

So far, we have talked only about humans or companies who could be responsible for Herzerg's death, but there is another possibility to consider. What if the AI itself is responsible? For simplicity, we will refer to the AI systems running Vasquez's autonomous Uber car collectively as 'the AI', even though it consists of many components. Standard cars are never morally or legally responsible for accidents they are in, even if we sometimes get angry at them when they fail to work properly. Does the intelligence of AI systems make them different?

One way to think about that question is to consider whether the AI could or should be subject to legal sanctions for harms it causes. It would not make any sense to try to impose some kinds of legal sanctions on this type of AI. For example, it would do no good to put Uber's AI hardware and software in a prison, nor would it be helpful for Herzberg's surviving family to try to collect financial damages from Uber's AI unless the AI was allowed to own money. It is difficult to see why or how AI could be liable to these legal sanctions if they are impossible to implement.

That said, the fact that some sanctions are impossible to implement now does not mean that the same sanctions will be impossible to implement in the future. One can imagine AIs of the future owning financial assets, for example those that are used for automatic stock trading. In such future cases, AIs could potentially be held legally responsible for harms they cause, by imposing fines on them or forcing them to give up some of their property. Some have also argued that some future AIs will have certain legal rights, like the right to go where they want or the right not be destroyed. If so, perhaps AIs could be legally punished by restricting their movements or even through physical destruction, akin to something like the death penalty. For now, though, Uber's self-driving AI systems do not seem susceptible or liable to legal sanctions, so they are not legally responsible.

What about moral responsibility? Are Uber's AI systems subject to moral sanctions, like moral guilt or moral outrage? Well, Uber's AI systems can't feel emotions, so they can't feel guilt, and it's not clear what it would mean to say an AI system should be built so that it has the ability to feel guilt. Many of us might be angry at the Uber self-driving car or its AI system for making a mistake that led to an unnecessary death, but such anger is misguided, even if understandable, if the self-driving car is no more responsible for what it does than an old-fashioned car is responsible for failing to start on a cold winter morning. Again, AIs of the future might have qualities that make them liable to moral guilt or outrage, but that's not a problem yet.

In contrast, other kinds of moral sanctions might seem appropriate for the self-driving cars of today. If your neighbour punched your friend because they supported a rival sports team, you might be justified in publicly condemning that neighbour as violent. You also might never talk to them again, which is a kind of ostracism. These kinds of moral sanctions can be applied to current self-driving cars as well. One could publicly condemn Uber's self-driving cars as dangerous and ostracize Uber's AI (in a way) by not using it until it is improved.[24]

One reason to publicly condemn and ostracize *people* who are morally liable is to make them feel morally guilty. Another reason is to prevent them from doing the same morally wrong thing again. Since Uber's AI could not hear us publicly condemning it and was not built to know or respond to the fact that we aren't using it, these outcomes of moral sanctions would not be achieved if they were applied to Uber's AI. Nonetheless, moral condemnation and ostracism can still serve other purposes, such as warning other people to avoid Uber's AI systems or take action to prevent them from being allowed on public roads in their current form. Thus, the fact that moral condemnation and ostracism would not make Vasquez's self-driving car feel bad is not enough to show that these sanctions do not serve at least one of the same purposes – harm prevention – in AIs as they serve in humans.

Where does this leave us? Was Uber's AI morally responsible? We believe that in some limited ways it can be. Its moral responsibility can justify some sanctions, including condemnation and ostracism, but not others, like imprisonment for malicious acts. Humans who control or influence the use of AI-driven cars are *more* morally responsible for the cars' actions than Uber's AI and liable to additional kinds of sanctions. However, those differences in degree and kind between the responsibility of humans and AIs are compatible with recognizing that the AI can be responsible to some degree in some ways.

Admittedly, if Uber's AI is responsible even a little, it is not easy to say exactly why the same arguments don't also apply to a faulty toaster oven that keeps causing fires, and yet most people agree that the toaster oven is not morally responsible at all. Of course, a toaster oven without AI does not gather information about its environment or learn the best means to achieve a goal (perfectly browned toast!). In contrast, an AI controlling a self-driving car can adjust its path and speed in light of information it senses in the present and has gathered in the past. Despite these differences, you may still think it is preposterous to ascribe any moral responsibility at all to a self-driving car. That's OK. The important thing for our current purposes is that you understand why some people might conclude that the AI of today can be morally responsible in some ways and that AIs of the future could be even more legally or morally responsible.

Responsibility gaps

The fact that society is still figuring out who is responsible for harm in cases like these can lead to the result that nobody is held responsible in the meantime, even if we all agree that somebody is responsible. These 'responsibility gaps' remove important incentives for people to use AI with due care and make it difficult to compensate victims of AI harms. The more pronounced these types of responsibility gap are, the less likely it is that people take sufficient action to avoid AI harms and the less recourse victims of AI have.

We have focused on one specific case in this chapter, but AI responsibility gaps can arise in most other uses of AI as well. For example, if hospitals use AI to detect cancer and the AI makes diagnostic mistakes that lead to some patients receiving unnecessary treatments or treatments too late, who is responsible for these misdiagnoses? Who, if anyone, should be forced to pay damages? What about if military units use AI to guide missiles and drones, but an AI-driven weapon kills an innocent family instead of the targeted terrorist? Who is responsible for the civilian losses? If no human is held responsible because no human knew what the AI would do, will military leaders have sufficient incentives to avoid such accidents? Will they get away with murder? When open-source chatbots give false medical advice that causes citizens to take poisons they believe are effective weight-loss treatments, who should be punished and who should help pay for the resulting medical care? When companies use AI to gather private financial and medical information about their customers in order to target their advertisements and coupons more effectively, such as when Target's coupons revealed a 16-year-old's pregnancy to her father, who is responsible for this invasion of privacy? Likewise, when social media AI algorithms facilitate hate crimes, as discussed in previous chapters, who should be held liable?

Our goal in this chapter was not to answer all these questions but instead to show how challenging they will be to answer, especially when the true future benefits and harms of an AI system in development are unknown, and when increasing networks of entities are involved in specific AI deployments. If we don't yet know who is responsible for AI's harms, it's hard to imagine that any effective plans can be put in place to mitigate such AI-related harms in the future. At the very least, any plans that are put in place will not be as effective as they could be unless we figure out who or what is responsible.

This issue is even more concerning in the face of the current frantic race for new AI products in response to the release of large language models. Many technology leaders, such as Google, developed AI products years ago that they refrained from introducing to the public due to ethical concerns, but many companies no longer feel they can afford to keep some of these products off the market.[25] If these companies are not held responsible for harms their AIs cause, they will face increasing pressure to create lucrative AI products with questionable safety profiles, even if they originally had good intentions. The case of large language models is also interesting because of their general applicability and the ability of other companies to use them in their own AI products. If a new company creates a medical advice chatbot based on OpenAI's GPT models, and this chatbot ends up giving misleading and harmful advice, is the new company responsible for this or is OpenAI?

In addition, if AIs themselves come to be seen as having the potential to be held legally or morally responsible for harms they cause, their creators will have more incentive and ability to deny their own responsibility. The result would be a world where AI creators can avoid negative sanctions by displacing their responsibility on to their products. This would dramatically diminish the options society has for using policy and legal recourse to minimize AI's harms. Thus, we need to think carefully about whether and when AI itself can be responsible and should be held responsible for certain actions.[26]

The appropriate ascriptions of responsibility and targets of sanctions will and should be debated as we gain more experience with AI systems. Societies will have to try different ways to distribute responsibility and sanctions in order to discover which legal and moral systems can help us achieve the benefits and avoid the dangers of AI systems among us. We need a balanced approach that minimizes unacceptable harm and incentivizes improvement without stifling innovation. Like Goldilocks, we want our rules of responsibility to be not too hot, not too cold, but just right. While we are figuring out what those responsibility rules are, though, we must be careful not to leave AI's harms unchecked, especially given the pace of progress. We will discuss ways to do that in the next two chapters.

CHAPTER 6

Can AI incorporate human morality?

A transplant surgeon told us a story about one night when a call woke him up. The caller informed him that a car accident had just killed an organ donor, so a kidney was available to transplant into one of his patients. The surgeon needed to leave immediately for the hospital, because the chances of a successful transplant went down by the hour. He also needed to tell the caller which of his patients to prepare for surgery. Unfortunately, the surgeon was still groggy from sleep, did not have quick access to his patients' records, and knew and liked some of his patients better than others. In these ways and others, his situation was far from ideal for making a life-changing moral decision about which patient should get the kidney.

Could AI have helped him? AIs do not get groggy, they do not forget details about patients, and they do not favour some patients over others merely because they like those patients' personalities. Moreover, they can make decisions like this one very quickly at any time of day. In these ways, a properly designed AI might be able to avoid some (though not all) of the problems that made our surgeon less likely to make the right choice. When we mentioned this possibility to the surgeon, he said he wished that this kind of help had been available.

This situation is just one among many where humans make important moral judgments in much less than ideal circumstances. Similar problems arise in decisions about which targets to attack in war, which criminal defendants to grant bail, whether to brake (and which way to turn) a car in an apparent emergency, and many other cases, as seen in Chapters 2–5. If AI could help humans make better moral judgments in even a few of these cases, it could benefit us tremendously.

Of course, this hope depends completely on whether the AI was designed well. An AI could also lead us far astray if it made inappropriate moral judgments, such as destroying humanity by maximizing paperclips in the horror story about the dangers of superintelligence that we discussed in Chapter 2.[1] The crucial question, then, is whether we can design AIs in such a way as to help us make better moral decisions without becoming too dangerous.

One proposal for doing this is to study human morality well enough to build that human morality – not a special AI morality – into AIs.[2] Humans would surely find it immoral for an AI to end humanity by producing too many paperclips. If AIs could learn which decisions are immoral according to human standards, then they could give us advice that might help us avoid moral mistakes in situations like the one described above with the surgeon. And if they were designed to avoid decisions that strike us as immoral, they would also be less likely to make such decisions. More specifically, we have discussed moral problems regarding AI safety, privacy, and fairness in this book. If AIs identify and avoid decisions that are not safe enough, that invade privacy without sufficient justification, or that are unfair, then their overall behaviour will be more ethical and will have impacts on society that are more consistent with our moral values.

Of course, the goal of building human morality into AI is not simple. The technical implementation of such a goal is also complex. There will be many practical, theoretical, and social challenges to navigate. Moreover, there will be various paths one might take to reach this goal. We will discuss a few options.

Morality from the top down?

One general approach to build human morality into AIs is to program moral principles with a high level of generality and then give the AIs guidelines about how to apply the principles to particular situations. This approach is intuitively attractive, because it allows us to be in control of which moral standards we give AIs. The challenge for this approach is probably immediately apparent, though: which moral principles should we program in?

Among the popular contenders are Isaac Asimov's three laws of robotics, which we will generalize to AIs of all kinds (not just robots):[3]

(I)   An AI may not injure a human being, or, through inaction, allow a human being to come to harm.

(II) An AI must obey the orders given it by human beings except where such orders would conflict with the First Law.

(III) An AI must protect its own existence as long as such protection does not conflict with the First or Second Laws.

These simple rules might be plausible starting points, but they quickly run into trouble. Is it true that we really want AIs to prevent *any* harm to *any* human, especially emotional harms? What if an AI removes offensive social media posts, but the authors of the posts become emotionally distraught that their content was taken down? What should an AI-driven robot do when one human attacks another human, and the robot cannot prevent harm to the victim without injuring the aggressor? Asimov's laws are inadequate to handle these types of situations.

Perhaps philosophers can do better. The problem with deferring to them, however, is that they disagree about which moral principles or theory should be built into an AI. Each moral theory has its advantages and disadvantages. For example, one moral theory called consequentialism proposes that we should do whatever maximizes good consequences and minimizes bad consequences in the long run. This sounds like a good thing to do. However, to apply consequentialism to a specific decision, the decision-maker has to have information about how all their possible actions could affect all beings in both the short and long term. Nobody knows how to obtain such an immense amount of information with enough certainty, especially in a computationally tractable way. Further, consequentialism applied the wrong way could lead to scary consequences. For example, a consequentialist AI that was programmed to 'minimize the total amount of suffering and death in the long run' might determine that it should kill all humans now, since that would prevent all human suffering in the future.

Other types of moral theories propose certain moral

rules, like 'Don't lie', or basic moral duties, like duties of fidelity, reparation, gratitude, beneficence, non-maleficence, and sometimes also justice and self-improvement. It might be more practically and computationally feasible to design an algorithm to follow these rules or fulfil these duties, since doing so does not require processing information about all possible consequences of each act. However, it is still an open question how we justify picking one set of rules and duties over another.

Further, these moral theories are difficult to implement in the context of AI. Most people think moral rules should be broken sometimes, for example in a situation where breaking the rule 'Don't lie' is the only way to prevent an assassin from killing your friend.[4] What should we tell an AI about which reasons are adequate to override a duty or rule? We could attempt to write exceptions into the AI's moral reasons, but the number of exceptions we would need seems endless. Duties and rules can also sometimes conflict with each other, such as when a surgeon has duties to two patients who both need the only available kidney, or when a judge cannot deny bail to reduce risks of crimes without harming the defendant's innocent family. What should we tell the AI about how to prioritize duties and rules? Without some scalable way of resolving conflicts between overridable rules and duties in novel situations, an AI will not be able to reach a conclusion that will reliably be acceptable to us.

Morality from the bottom up?

To avoid these challenges, another approach might try to avoid general rules and instead enable AIs to learn human morality from lots of specific examples of behaviour and de-cisions that humans judge to be morally good or bad. An AI of this kind wouldn't need to endorse or even understand the moral judgments, but it would need to be able to accurately predict which moral judgments humans would endorse in a certain situation or dilemma. We will call this the 'bottom-up approach'.

This method has at least one strong advantage. Whereas the top-down approach begins with a moral principle that some people are bound to question, AI creators using the bottom-up method do not tell the AI which categories or ar-guments are relevant for predicting human moral judgments. The AI is supposed to figure that out on its own. That min-imizes a lot of the decisions AI creators would otherwise have to make about which principles and rules to follow and how to adjudicate between them in individual scenarios. In this way, this approach sounds easier to implement than the top-down approach.

However, the bottom-up method also introduces its own problems. First, the bottom-up method would require train-ing the AI on a tremendous amount of very diverse data, given that the AI would have to learn from many different types of moral situations. If the data used to train the AI was biased or limited in key ways, it might make incorrect predictions in cases not well represented in the training data or when trying to predict the moral judgments of people who make very different judgments than those in the training data. Second, it's not clear which kinds of data to train the AI on. Written descriptions of what it is right or wrong to do in a specific scenario? Questionnaire ratings? Actual human choices and behaviours? Something else? Third, how can we determine whether *all* relevant data has been included in training an AI? Can any of these methods of eliciting people's moral opinions adequately capture all features of the scenarios that affect our moral opinions in real life? Fourth, what if some people have moral opinions that others find reprehensible? And what if individuals change their own moral opinions over time, when they are in certain moods, when they are tired, or when the same situation is described in different ways?[5] Do such unstable opinions reflect their basic values? Fifth, a bottom-up system might be able to predict *which* acts humans will judge as wrong, but it would be less likely to be able to give us substantive reasons *why* humans judge those acts and not others as wrong, or why the AI predicts they are wrong. This relative opacity would make it hard to anticipate when the AI is likely to make mistakes or go astray, and hard to correct these mistakes. These issues make it difficult to see how a bottom-up method could succeed.

The best of both worlds

The problems plaguing top-down and bottom-up methods leave us wanting a better alternative. We believe that the most promising strategies for building morality into AIs will combine top-down and bottom-up approaches.

These hybrid methods will not be simple to develop, but we have started testing one approach in a very constrained decision-making context within a specific application of AI. Limiting the scope of the moral judgments that the AI needs to make and the moral principles or patterns that the AI needs to learn makes the myriad of technical challenges a little bit more tractable. As we build up knowledge of how an AI could be used for these constrained moral judgments, we might be able to leverage those lessons to expand the strategy to wider moral contexts. But we begin with the concrete test case of kidney allocation under scarcity.

Who gets the kidney?

Roughly 100,000 people in the US alone are waiting for kidney transplants.[6] There are two main sources of kidneys. A kidney transplant can come from a cadaver, such as someone who died in an accident very recently. Alternatively it could come from a living donor, since most of us have two kidneys but only need one to live, so we can choose to donate one of our kidneys to someone who needs it.

Not all donor kidneys are compatible with all possible recipients. Donors and recipients have to have compatible blood types and immune profiles for a transplant to be viable. Further, compatibility is on a spectrum, and some pairs of blood types and immune profiles can be made more compatible – meaning that the kidney will function successfully for a longer amount of time – if the recipient takes immunosuppressant drugs.

When a kidney becomes available for transplant, it is often compatible with more than one needy recipient, and there are not enough donors (live or dead) to supply all patients in need. As a result, doctors or hospitals often have to decide which one of several needy patients should receive a kidney that becomes available. The patient who gets the kidney, assuming it is adequately compatible, will usually gain anywhere from ten to twenty extra years of life from the transplant.[7] A needy patient who does not get the kidney has to keep waiting, and may lose a lot of quality of life as their condition deteriorates. They may even die waiting. Thirteen kidney patients die each day before receiving a kidney transplant that would have been able to save their life.[8] So the decision of who gets a kidney that becomes available has true life-and-death consequences.

Kidney allocation decisions are made in different ways in different transplant centres, but AI tools are becoming increasingly available to make these decisions more efficient and easier. For example, AI systems have been used with great success to facilitate kidney exchanges in which patients who have willing but incompatible donors swap donors with other patients in the same situation.[9] In addition, AI decision-aids that predict the compatibility of a kidney with a patient, or that predict the likelihood that a more compatible kidney will become available in a certain amount of time, are being tested.[10]

Even though these AI tools tackle different aspects of the kidney allocation process, a critical question for all of them is which features they should take into consideration when making decisions. Currently, most kidney transplant decisions are made on the basis of medical compatibility, age, health, organ quality, and the patient's time on the waiting list. These features are typically considered objective medical or pragmatic factors, and they are prioritized for that reason. However, citizens outside kidney policy-making teams think that additional features of kidney patients should also be taken into account when deciding who gets a kidney, often for moral reasons. For example, many people believe that the number of dependants, violent crime record, and behavioural choices that might have worsened the kidney disease (like smoking) should impact who gets a kidney. Many also want to make sure that race, gender, and religion are *not* taken into account at all when deciding who gets a kidney.[11]

How can moral judgments like these be built into AI tools to help with kidney allocation? We focus on two paradigm uses of AI. The first is a decision-aid to help transplant surgeons decide whether or not to accept a kidney on behalf of a patient. Sometimes surgeons have to make such decisions under extreme time pressure or when they are groggy after having been woken up in the middle of the night (as in the case that opened this chapter). An AI decision-aid would learn how an individual surgeon considers medical and moral issues in kidney allocation decisions when that doctor is in a well-rested, calm, and better-informed state. Then, when the surgeon has to make a decision in suboptimal conditions, the AI would tell them what it predicts they would decide in better circumstances, so that they can take that prediction into account when making their final decision. This same tool could potentially be used as a stand-in for a transplant surgeon if they were unreachable when a kidney decision needs to be made quickly.

The second use of AI involves building the collective moral judgments of a hospital's transplant community – which typically includes doctors, nurses, administrators, lawyers, other experts, patients, and sometimes even lay community members – into automated systems, which some institutions already employ, to determine which patients will be offered a kidney and in what order. The goal is to make the results of the automated kidney allocation priority lists align with the community's moral values.

These two types of tools each need to solve some unique problems, like figuring out how to aggregate a group's diverging moral preferences in the second tool but not the first. Still, they both can be pursued with in the same core strategy. Here's what that core strategy looks like.

How to build morality into kidney allocation AIs

IDENTIFY MORALLY RELEVANT FEATURES

The first step of our strategy uses open-ended surveys to crowdsource judgments about which features of patients should and should not influence who gets a kidney. These responses are analysed, refined, supplemented, and consolidated into lists. New survey participants then confirm whether or not each listed feature is indeed morally relevant to kidney allocations.

Our surveys ask for judgments about what *should* and what *should not* influence who gets a kidney. The first provides morally relevant features that AIs should use to predict human moral judgments. The second indicates which features should intentionally be ignored by the AIs when making decisions.

The survey participants comprise a broad group of stakeholders, including the general public, kidney patients and their families, members of demographic groups most likely to contract kidney disease, doctors, nurses, hospital administrators, and ethicists. This wide range of respondents is necessary to ensure both respect for expertise and inclusion of affected groups so that there will be, as the slogan goes, 'Nothing About Us Without Us'.

These broad surveys are not meant to determine which patient features really should or should not influence who gets a kidney. No mere survey can tell us what really is morally right or wrong. Instead, crowdsourcing is meant to construct a list of moral features that people see as morally relevant, so kidney allocation AIs can use these features to predict people's moral judgments accurately, and people will be able to understand the basis of the AIs' predictions.

The result of this crowdsourcing is three lists: features that are generally seen as morally relevant, features that are generally seen as morally irrelevant, and features that are controversial. In our studies so far, the vast majority of participants agreed that race, gender, sexual orientation, religion, political beliefs, wealth, and reliance on government assistance should *not* influence who gets a kidney.[12] It is also not surprising that almost all of our participants agreed that urgency of need, time on the waiting list, and likelihood of transplant success *should* affect who gets a kidney, along with age, current health, life expectancy, and quality of life after a transplant. Most also said that smoking and drug and alcohol abuse (either before or after diagnosis with kidney disease) should matter. It was more controversial whether mental health, record of violent or non-violent crime, or number of children or elderly dependants should affect who gets a kidney. Although the rationales for many of these features need to be clarified through further investigation (e.g. do people just use age and current health as proxies for life expectancy?), we can use these preliminary lists to start figuring out how to incorporate human moral values into AI systems.

MEASURE MORAL WEIGHTS

After constructing a list of morally relevant features, we need to determine how people incorporate these general features into moral judgments about specific kidney allocations. We want to know how much weight people put on each feature, as well as how the presence or absence of other features can affect that weight. For example, perhaps you think that patients who don't smoke should be strongly prioritized above those who do, but only if the patients have children; when the patients don't have children, you think smoking shouldn't matter much, perhaps because their second-hand smoke isn't affecting any children within their home. We also need to know how people resolve conflicts. For example, if you think patients who don't smoke should be strongly prioritized and younger patients should be strongly prioritized, should a young smoker get the kidney instead of an older non-smoker?

In an ideal world, we could simply ask participants how important each feature is in different contexts and how to resolve conflicts. Unfortunately, people aren't very good at answering those types of questions. Thus, a major area of research – sometimes called 'preference elicitation' – is figuring out which methods best allow people to share their moral judgments about specific issues.

Even though the jury is still out about the best method, here is one we have used with some success. We tell people that an available kidney can be given to one of two patients,

Patient A and Patient B. We provide information about key features from our crowdsourced list for both patients. For example, a simplified version might say that Patient A will be physically able to work 30 hours a week after receiving a kidney, is underweight, will gain five additional years of life by getting the transplant, has one elderly dependant, and has been waiting for a kidney for five years. Patient B will not be physically able to work even after receiving a kidney, is obese, will gain ten additional years of life by getting the transplant, has no elderly dependants, and has been waiting for a kidney for seven years. Then we ask participants, 'Should the kidney go to Patient A or Patient B?' We can also give participants the option of choosing Patient A or Patient B randomly (by flipping a coin).[13] We ask participants to make this decision many times, each time varying the difference between the values in different features and each time creating different conflicts between features. We then use these decisions to teach an AI how much weight participants put on certain features of patients, as well as how those features interact in producing their moral judgments about who should get the kidney.

One obvious challenge for this approach is that the number of comparisons we need to ask participants to make judgments about grows exponentially with the quantity of features and their possible values for each feature. It quickly becomes infeasible to ask participants about all possible conflicts and computationally intractable to determine all possible ways in which features interact. Nonetheless, we can still make progress by focusing on a limited number of features at a time and by leveraging technical tricks that reduce the number of comparisons that need to be asked. One trick, called 'active' learning, allows us to adjust the comparisons and conflicts in real time so that we ultimately ask the subset of questions that will be most informative for learning how a specific individual makes their moral judgments.

MODEL MORAL JUDGMENTS

Once we have collected enough data about how people weigh features, we statistically model people's moral judgments. We typically apply an AI that employs a bottom-up approach to learn: (A) which features really do influence participants' moral judgments about who should get a kidney; (B) how these features interact to produce an overall judgment; and (C) which models best predict the moral judgments of individuals. In this way, AIs can learn how to predict human moral judgments about allocating kidneys.

A core tenet of our strategy is that we exclusively employ interpretable AI methods. For us, a method is sufficiently interpretable only if it allows us to identify which features affect people's moral judgments and to understand their influence. This level of transparency is important for at least five reasons. First, it allows us to be confident that the features used by the AI align with features that humans see as morally relevant. Second, if AI predictions of human moral judgments are used in real settings, such as kidney allocation, stakeholders deserve to know how the AI system addresses moral issues. Third, knowing how the AI uses relevant features makes it easier to ask humans for feedback about whether the AI is predicting accurately, which can be used to improve the system down the line. Fourth, if the AI uses features and weights that humans recognize and understand, it can reveal not only which moral judgments a person will make but also potentially *why* they make those moral judgments – that is, their reasons for those judgments. Finally, all of this will help patients, doctors, and communities have more trust in the AI system, which will in turn make them more willing to work with it and use it. Transparency enables trust, which is necessary for wide adoption of the AI system, along with all of its potential social benefits.

AGGREGATE GROUP JUDGMENTS

So far, we have discussed how AI can predict individual human moral judgments. These predictions are the end goal of some useful AI systems, such as decision-aids that predict which kidney transplant recipient a specific calm and rested surgeon would choose. For other applications, such as when we want to build the collective moral judgments of a hospital's transplant community into its automated patient priority systems, we need to predict the moral judgments of human groups. Applications of AI in these areas require individual human moral judgments to be aggregated in some way.

A simple aggregation strategy is to count the members of the community that we predict would favour each potential recipient of a kidney: let the majority rule! We could also weight the judgments of certain stakeholders more heavily, such as those with more expertise or more at stake. Much more sophisticated aggregation schemes draw on social choice theory, which is a framework for analysing different methods of combining individual opinions into collective decisions. An active research community is dedicated to developing these techniques for cases where we lack perfect information about each individual's judgments.[14] Here we will not endorse any particular approach to aggregation but will simply note that an approach to this problem needs to be chosen before AI can be used to predict moral judgments by groups.

IDEALIZE MORAL JUDGMENTS

So far, we have focused on predicting which moral judgments real people make in real circumstances. This is sufficient for building some important aspects of human morality into some AI systems. However, making some AI systems behave morally will require addressing an additional problem: AIs that make the same moral judgments as humans will also make the same moral *mistakes* as humans.

We all make a lot of moral mistakes, even by our own standards. We sometimes forget or fail to attend adequately to facts that we ourselves see as morally relevant. When many complex considerations support each side of an issue, we can get overwhelmed and confused. In addition, intense anger, disgust, and fear often lead us to base our moral judgments on factors that we think should not matter (after we calm down). Yet another source of moral error is bias in a broad sense, which includes cognitive biases, favouritism, and racial or gender prejudices that we consciously reject but that nevertheless unconsciously impact our decision-making. In sum, we can all be misinformed, forgetful, confused, emotional, or biased when we make moral judgments. If AIs are trained on these judgments, the AIs will reflect and perpetuate the results of our misinformation, forgetfulness, confusion, emotion, and bias.

Despite our imperfections, most of us want our own moral

judgments and decisions to be less subject to such distorting influences. We also often want the morality built into an AI to reflect the moral judgments that we ourselves would make if we were more informed, rational, and unbiased, even if we will never actually be in an ideal state. In such cases, we want AIs to predict and reflect our *idealized* human moral judgments, not our actual moral judgments.

Idealizing will be one of moral AI's greatest challenges, but it will also provide one of moral AI's greatest advantages if the challenges are overcome. Despite centuries of trying, we still have no way of preventing humans from making the moral mistakes that an idealized moral AI would be able to avoid. Further, if we can figure out how to correct for biases or mistakes in a moral AI's training data and models, the AI's predictions and actions would align better with our own fundamental moral values. They might also be able to advise *us humans* about how to make actual judgments that are better aligned with our moral values.

So how do we idealize a moral AI? The first thought that might come to mind is to try to find a data set of ideal human judgments to train the AI on. The problem with this proposal, of course, is that our human limitations prevent us from knowing which human judgments are ideal. Thus, we need some other way to know or predict what idealized moral judgments would look like. At present, it may be impossible to know or predict perfectly idealized judgments, but we can still make progress towards this goal by combining corrections for individual sources of moral mistakes, bias, or confusion. Let us explain what we mean with an analogy.

US Supreme Court Justice Thurgood Marshall stated in his judicial opinions about capital punishment:

> In *Furman* [*v. Georgia*, 1972], I observed that the American people are largely unaware of the information critical to a judgment on the morality of the death penalty, and concluded that if they were better informed they would consider it shocking, unjust, and unacceptable.[15]

A study confirmed that, indeed, the more people learned about critical facts related to the death penalty, the more opposed to the death penalty they were likely to be.[16]

Analogously, additional information might also affect our moral judgments about kidney allocations. Most of us do not know much about what life is like for a patient on dialysis, how likely a kidney transplant is to succeed and which complications and side effects are common after one, or whether smoking tobacco or drinking too much alcohol exacerbates chronic kidney disease.[17] Coming to know this information could change our moral judgments about which factors should influence who should get a kidney, and in what ways. If so, moral judgments made without access to this information could be said to be plagued by mistakes due to ignorance.

Fortunately, if we can measure the impact of ignorance on moral judgments about kidney allocations, we can correct for it in an AI system. Here's how we do that. First, we ask people for their judgments about who should get a kidney using the methods we described earlier. Next, we use surveys to ask participants how much they know about specific types of information relevant to kidney disease and allocation, and we confirm their knowledge with assessments. Then, we teach participants about some of the information they didn't know, and we do so to varying degrees with different people. Finally, we again ask people for their judgments about who should get a kidney. By comparing the change in moral judgments before and after teaching sessions to the moral judgments of people who weren't given a teaching intervention, we can estimate the impact of additional information on moral judgments about kidney allocations. This, in turn, will allow us to predict what the moral judgments of similar people would be if they had adequate information. Then we can mathematically correct for this effect in kidney allocation AI tools.

Accounting for ignorance is one kind of correction we would want to make, but most people also want their moral judgments to be free from bias. When we say that a moral judgment is biased, we mean the judgment results from some feature(s) that we believe should not affect the moral judgment. For instance, most participants in our surveys agree that race, religion, gender, sexual orientation, wealth, and attractiveness should have no impact on who receives a kidney. Nonetheless, most of us have biases associated with these attributes, even if those biases are implicit or unconscious; and these biases can affect moral judgments about kidney allocations. Can an AI correct for these biases, too?

One first-line strategy to reduce biases is to hide any information that is likely to bias moral judgments directly when collecting the training data for an AI. When asking whether Patient A or Patient B should get the kidney, for example, we can simply not tell participants anything about the race, gender, religion, sexual orientation, wealth, or attractiveness of either patient. This 'veil of ignorance'[18] will reduce the likelihood that unwanted factors influence participants' moral judgments directly.

Unfortunately, biases can still wriggle their way into our moral judgments *indirectly* through proxy variables (as discussed in Chapter 4). For instance, even if, without mentioning race or gender, we tell you that Patient A has a criminal record and Patient B does not, but you associate criminal records with a certain race or gender, then your attitudes towards that race or gender can indirectly affect your moral judgment of whether Patient A or B should get the only available kidney, even if you are not aware of it.

AI could potentially correct for these kinds of biases using a similar approach to the one we described for correcting ignorance. This time, though, instead of modelling the impact on moral judgments of teaching participants varying amounts of background information about kidney allocations, we model the results of including factors for which we want to correct bias (such as race). To do this, we ask participants judge who should get a kidney, using sets of questions that allow us to measure attitudes to those factors.

Suppose we give survey participants multiple scenarios in which Patients A and B differ in whether one, both, or neither patient has a criminal record and whether one, both, or neither patient is Black. We also include some scenarios where either criminal record or race is not mentioned. Most of the scenarios also describe other patient features, such as time on waiting list, number of dependents, and so on; and we can hold those other features constant when necessary.

As long as we ask patients appropriate scenarios presented

in interleaved orders, we can analyze participants' responses to determine whether, how much, and in what direction the race and criminal record of a patient influences participants' judgments about who should get an available kidney. Using the right set of scenarios and analyses, we can also determine whether participants are associating criminal records with being Black so that they are treating criminal record as a proxy for race (or vice versa). The same methods can be repeated for other suspected proxies.

With these influences and associations figured out, we can model the direction and magnitude of racial bias in different settings, both directly and through proxies like criminal record. Further, we can use the model to mathematically extrapolate the racial bias that we can expect to see in scenarios that have different feature values than were in the training data. Then we can use these estimates to mitigate racial bias built into a moral AI that was trained on participants' moral judgments.

Ignorance and bias are just two ways in which human moral judgments can fail to be ideal, and there are plenty of further technical questions about precisely how to conduct the analyses and extrapolations. Still, hopefully you get the general idea of how idealization could be pursued. Nothing, in principle, stands in the way of gathering the kind of information that an AI would need in order to predict the moral judgments that humans would make if they were in more ideal circumstances in terms of being more informed, rational, and impartial or unbiased. It will just take time to figure out the precise experimental methods needed to make all the specific kinds of corrections we might want to make.

This idealization step needs to be carried out very carefully because our best models of human moral mistakes will never be perfect, so any corrections based on these models could end up introducing their own source of mistakes. Even when individual implementations of this method succeed, we are not so naïve as to think that it will ever lead to flawless moral machines. Still, we believe that building idealized moral judgments into AIs could be a big improvement over AI technology that has no morality or uncorrected actual morality built into it.

Artificial improved democracy

Our group has focused on developing AI to guide moral judgments about kidney allocations. Our goal in doing this is *not* to create an AI to tell people what is *really* and *truly* moral or immoral when allocating kidneys. Nonetheless, an AI that is trained and corrected as we described above could still be helpful for providing evidence that will help us decide what we ought to believe and do in complex moral situations.[19] Further, an AI that has morality built into it through our method will be more likely to behave in ways we find morally acceptable than an AI that does not.

Our goal is also *not* to replace doctors. Whenever AI is used in medical contexts, human medical professionals should have the final say about what to do. We only want to help doctors and others make those decisions with more information and with appropriate levels of humility and confidence. We hope that doctors who must make difficult moral decisions without adequate information under time pressure – and perhaps while groggy, like the surgeon in the story that opened this chapter – will be grateful for such help.

Despite these qualifications, our broader goal is to extend this use of AI into many other areas. Our overall method for building morality into AIs can be understood as *artificial improved democracy* (AID). It is *democracy* insofar as it rests on moral judgments from the general public. As in elections, the process can involve preferences, values, and moral judgments of all stakeholders. Our method *improves* on democracy by correcting for common errors – like ignorance, confusion, and bias – to reveal what they would judge if they were more informed, rational, and impartial than real life typically allows. The point is not to bypass people's preferences or values but rather to find out what they really want and judge to be morally right and wrong. Our method is *artificial* because it is manifested through artificial intelligence. Machine learning enables predictions in new cases without needing to hold a separate referendum on every decision. Overall, the goal of our method is to *aid* people and AI systems in making better moral judgments and behaving in ways that are more in line with human moral values.

When built into decision-assistance tools, AID could potentially help people avoid the most common sources of error in human moral judgment in a wide range of areas. It could help job recruiters avoid bias in their decisions to interview or hire job applicants. It could help military operators avoid immoral decisions about when to fire missiles at which targets. It could help social media content moderators be more consistent.

AID can also help to prevent AIs themselves from threatening safety, privacy, and justice in the manner discussed in earlier chapters. Of course, building and implementing these systems will require a lot of hard work, and it will take time to figure out how to elicit, model, and idealize every kind of moral judgment we might want to build into an AI. It will also take time to figure out how to make sure humans use AID in ways we find acceptable, without trusting an AID too much or too little. Still, in principle, we could use AID to stop an AI from doing any act that enough informed, rational, and impartial humans would judge to be morally wrong. That hope motivates our work.

What can we do?

In the previous chapter, we described a general strategy for building human morality into AI so that AIs will be more likely to behave in safe ways that align with our ethical values. This is one of many technical approaches we have shared throughout this book which can help make AI more moral. We are optimistic about the promise of these approaches and, as we have said, dedicate much of our own research to developing related technology.

At the same time, it is clear that technology, on its own, will not be enough to ensure that AI's impact on society is what we want and need it to be. Despite their mathematical and algorithmic underpinnings, AI systems are still created by humans, funded by humans, and function within human society (at least for the time being). That's what this chapter is about. What is the human side of what we need to do to make sure that AI's impacts on our lives are ethical?

We obviously aren't the first people to ask that question. Entire top-quality conferences have been dedicated to the ethical use of AI and AI's societal impacts.[1] Multiple governments and companies have also initiated conversations about how moral AI could be pursued. These efforts have led to *more than one hundred* separate documents that outline 'ethical' values or principles the AI industry should incorporate into its practices.[2] Since these principles are often described as professional 'ethics', we will use 'ethical' and 'moral' interchangeably in this chapter. These principles include statements like:

> Highly autonomous AI systems should be designed so that their goals and behaviors can be assured to align with human values throughout their operation.
> — Future of Life Institute

> AI systems should be understandable; People should be accountable for AI systems; AI systems should treat all people fairly.
> — Microsoft's Responsible AI principles

> The data, system and AI business models should be transparent.
> — European Commission Ethics Guidelines for Trustworthy Artificial Intelligence

> There should be transparency and responsible disclosure to ensure people know when they are being significantly impacted by an AI system, and can find out when an AI system is engaging with them.
> — Australian Government AI Ethics Principles

> The department will take deliberate steps to minimize unintended bias in AI capabilities.
> — US Department of Defense Ethical AI Principles

> The development of AI should reflect diversity and inclusiveness, and be designed to benefit as many people

as possible, especially those who would otherwise be easily neglected or underrepresented in AI applications.

— Beijing AI Principles

We will incorporate our privacy principles in the development and use of our AI technologies. We will give opportunity for notice and consent, encourage architectures with privacy safeguards, and provide appropriate transparency and control over the use of data.

— Google's AI Principles

These commitments all sound really good, right? We agree! The problem is that despite such comforting mission statements and all the technical tools we've discussed in previous chapters, the ethical issues we described in this book persist.[3] Why?

One general explanation is that the AIs that impact our lives are created by NGOs (non-governmental organizations) that are not bound by moral AI principles, even if they help craft the principles and remain inspired by them. According to one survey in 2021, just 20 per cent of companies had an AI ethics framework in place, and only 35 per cent had plans to improve the governance of AI systems and processes.[4] Further, when technology experts were asked, 'By 2030, will most of the AI systems being used by organizations of all sorts employ ethical principles focused primarily on the public good?' a majority (68 per cent) said no. So there is widespread recognition of a disconnect between what moral AI principles have to offer and what is needed to create moral AI.

Another less appreciated reason is that even the best

moral AI technical tools require humans to make informed, thoughtful ethical decisions when implementing them, and most AI contributors have not had training in how to make those decisions. In illustration, software packages with names like IBM Fairness 360[5] and Fairlearn[6] help AI teams make their AI products fair by providing algorithms that mitigate bias using some of the approaches described in Chapter 4. But to apply these types of algorithms, the teams have to choose one of twenty different mathematical definitions of fairness, each with its own social, technical, or statistical limitations. How should they make the decision, especially if they haven't had the opportunity to learn about the societal impact of different types of fairness? No good strategy exists, which is why studies have confirmed that many teams, despite good intentions, do not end up implementing fair AI algorithms in ways that align with their goals.[7] In fact, non-experts often completely misinterpret AI fairness metrics,[8] so teams without deep technical expertise have even more trouble. This is just the tip of the iceberg of issues that separate moral AI technical tools from what is needed to make sure real AI products are used ethically.

The good news is that there are practical ways to fill in the moral AI 'theory-to-practice gap'. We need to be humble and clear-eyed as we pursue these opportunities, though, because ensuring all AI uses align with our values will be neither easy nor fast. We would love to provide a clear set of instructions that society and AI creators could follow to make AI ethical. Unfortunately, the way forward is going to be messier than that. AI is used in so many ways in such diverse aspects of life that it is impossible to create a single set of instructions that will adequately address all AI uses. Further, the contexts in which AI is used and in which AI has impacts are complicated, dynamic, and involve numerous human factors, so our approach needs to be flexible and responsive to unpredictable developments and evolving social contexts.

To achieve this, we think our moral AI strategy needs to be pursued on at least five separate battlegrounds in parallel to be successful: technology dissemination, organizational practices, education, civic participation, and public policy. To explain these strategic areas, first we need to introduce some important nuances in how AI is created.

Nuances of the AI creation process

AI CONTRIBUTORS OFTEN HAVE LIMITED COMMUNICATION

Recall from Chapter 1 that the AIs that impact our lives – like websites, phone apps, recommendations, robots, drones, and chatbots – are AI *products*. They are experiences, interfaces, or devices that typically allow us to interact with AI *models* that were trained for a specific purpose. The models are trained by iteratively passing carefully chosen data through an AI *algorithm* that outlines which mathematical functions will be applied to the data during training. When trained models and other components are combined together to serve as the backbone of an AI product, sometimes we call the combined entity an AI *system*. An AI system becomes an AI product when it is packaged (and usually sold) in a way that other people (often customers) can use it.

Also remember that the people who create AI algorithms, models, systems, and products tend to have very different backgrounds. AI algorithms are usually developed by researchers with PhDs in computer science, mathematics, or statistics, often in academic settings or in industry research labs. AI models are typically trained for specific purposes and put into systems by interdisciplinary teams of engineers and data scientists with diverse types of education in different settings ranging from corporations to classrooms or home offices, sometimes with organizations sponsoring the pursuit of those purposes. Those organizations can be companies, government agencies, research labs, or even educational institutions. In large organizations that have advanced AI capabilities, sometimes AI algorithm developers may work closely with the people training AI models and building the other technical components of AI systems. However, in more standard situations, engineers and data scientists use software packages that AI algorithm developers make available, but do not interact with those developers directly.

Yet additional skills and processes are needed to take the next step and turn AI models or systems into AI products. These additional skills are usually provided by product managers (or 'owners'), user interface researchers and designers, front-end software engineers, other types of engineers or data scientists who create technical infrastructure to run the product efficiently, and sometimes data scientists who help analyse user research. These team members may or may not have had experience analysing data using statistics or AI, and in many cases do not have graduate degrees in technical fields (although many software engineers and data scientists do).

Why are we reviewing all these different roles in a 'what can we do about it' chapter? Because one of the most critical issues moral AI strategies must address is that the different types of AI contributors described above often do not have the opportunity to communicate during the creation of a specific AI product. They might not even know about each other's existence. In particular, the people who create AI algorithms themselves – especially cutting-edge algorithms that try to address ethical concerns, like the 'fair AI' algorithms or 'interpretable AI' algorithms we have mentioned – are usually *not* part of an AI product team or even part of a product's sponsoring organization at all. Some companies with advanced AI capabilities – like OpenAI, Meta, Google, Microsoft, Amazon, Apple, or Anthropic – may make concerted efforts to facilitate close collaborations between AI algorithm creators, AI model trainers, and AI product teams, but such collaboration is often difficult to maintain and may be completely impossible to mimic in companies with different access to resources.

This disconnect will only become more prevalent as the 'AI as a service' (AIaaS) ecosystem we discussed in Chapter 3 grows. For example, consider the collection of large language models, like ChatGPT, that have appeared on the market. Even though companies of all kinds are launching an ever-increasing number of products with ChatGPT integrated into them, the people who create and refine ChatGPT's models are employed by the company OpenAI, not the companies creating the integrated products. Open AI may try to make their own engineers and model specialists available for consultation to their major corporate partners, like the Microsoft Bing team, but that won't be possible with all partners or with all large language models.

Even when AI contributors from different parts of the AI creation process do interact, their backgrounds and professional vocabularies are sometimes so different that many problems arise. It can be hard to figure out what information to share. Efforts to share information are fraught with misunderstandings. Group members are often unaware of what information is lost during translation. And so on. In sum, getting all the information relevant to an AI product's ethical impacts into the hands and brains of the people who need to act on the information is an immense challenge. This problem becomes even more formidable when teams are expected to work on tight timelines. We'll explain that issue next.

'FAIL FAST AND FAIL OFTEN' PERMEATES AI PRODUCT CULTURE

The field of AI product development has a dirty secret: most AI initiatives fail. According to one report, eight out of ten AI projects do not succeed or provide value, typically because they can't find enough appropriate data to train models, can't generate accurate enough models, are too expensive to train, or address a problem that users don't end up caring about.[9] The odds are definitely stacked against AI teams.

Despite all this, the AI product landscape is fiercely competitive. Corporations are racing to be the first to solve new problems with AI products, and most invest in multiple AI projects at the same time. As a result, the AI marketplace continues to expand. Venture funding for AI companies surpassed US$61 billion from 2010 through the first quarter of 2020,[10] and over $20 billion during just the first three quarters of 2022.[11] Thus, AI product teams have a lot of pressure on them to create financially lucrative AI products before their competitors do.

Within this challenging and competitive culture, AI product teams often have to navigate a mismatch between the inherent uncertainty associated with AI projects and what organizational leaders have come to expect from traditional *software* products, like apps. This mismatch has profound implications for whether moral AI will be given the organizational support it needs, so it is worth explaining.

In brief, software product teams often strive to be 'lean' and 'agile'. That doesn't mean that they strive to be physically fit but, rather, that they intend to 'fail fast and fail often'.[12] The motivating idea is that it is impossible to know fully in advance what makes a product great. According to 'lean-agile' thinking,[13] therefore, we should abandon strategies that attempt to make the best product imaginable before showing it to people. Instead, we should 'do the simplest thing that could possibly work' by relentlessly repeating the following cycle: invest as few resources as possible into a minimum viable product, ask people to use it, then incorporate feedback into the next improved version of the prototype. This approach has been proven to allow teams to arrive at successful products faster and with less resource investment than if they try to get everything right the first time around,[14] so it has become the predominant methodology for creating products that use digital technology.

Many lean-agile teams create prototypes using short time periods called 'sprints', during which a team agrees to complete a certain amount of work which culminates in a new product iteration or feature. Product teams aim to calibrate the amount of work they take on during a sprint so that the timeline is very challenging but still feasible. To meet such ambitious timelines, work is often delegated to separate groups of contributors according to their expertise, tasks are pursued in parallel, and team contributors are expected to stay focused on their assigned duties until the end of the sprint, when everybody's finished work is shared and evaluated.

Throughout this process, teams constantly monitor and aim to improve their efficiency by leveraging carefully designed 'metrics'. Metrics are quantitative measurements organizations track to evaluate the progress of a product or initiative. Some metrics, such as average percentage of tasks completed, are meant to track the team's processes. Other metrics, such as number of users gained, are meant to track the product iteration's reception. Especially because product teams' financial compensation is often tied to their ability to obtain certain metric values, if a phenomenon is not being tracked by a metric, it is typically not a major consideration in a product's development. The significance of this will become clear later.

Lean-agile methodology is pervasive in the tech world, with 94 per cent of companies reporting using such methods.[15] The problem is that unique aspects of AI products cause friction with standard lean-agile methods. First, lean-agile methods were honed in the software industry where, traditionally, timelines are fairly stable and predictable. AI development timelines, in contrast, are notoriously *unpredictable* due to the aforementioned challenges with wrangling appropriate data and generating acceptably accurate AI models.[16] The second issue is that it is often not clear what level of accuracy, fairness, or transparency is necessary for a minimum viable AI product. It may be fine to deploy a prototype of a wine-recommendation AI which is 70 per cent accurate, but most people would reject a self-driving car that stops at only 99 per cent of stop signs, ignoring the other 1 per cent. Further, as we discussed in Chapter 3, some dangers of AI products are not obvious, which makes it hard to tell when a prototype is appropriate for use in public. Some organizations have started to adapt their AI work methods accordingly, but many have not. In such situations, there can be intense tension between AI product teams and other parts of an organization that are organized around lean-agile timelines and metrics. This tension underlies many dynamics moral AI strategies need to tackle.

Why does the moral AI principle-to-practice gap arise?

With that crash course in AI development in our pocket, we are ready to begin dissecting the reasons moral AI technology and principles remain so detached from the AI products that actually get made. Many of the issues may already be apparent. Overall, moral AI technical tools and principles do not sufficiently address the ethical challenges that arise from the cultural, operational, and financial pressures on organizations making AI products, or the lack of communication opportunities between AI contributors or between AI contributors and AI users. Here are just a few examples of those unaddressed challenges.

DISEMPOWERING ETHICS

The German sociologist Ulrich Beck wrote that ethics often 'plays the role of a bicycle brake on an intercontinental airplane'.[17] Many organizations that try to develop AI products function in ways that make this description appropriate. AI teams repeatedly report that their organizations' financial requirements, timelines, expectations, and compensation structures are not compatible with the investment necessary to grapple with the ethical problems their AI products pose.[18] Even if AI contributors know how they could incorporate moral considerations into their products, and believe they should do so, they may not be empowered to take relevant actions, especially if an organization's leadership has not made clear how ethical concerns should be prioritized relative to other goals.

This disempowerment can be formal (through organizational charts that prevent contributors from influencing decisions) or informal (through priority suggestions that consistently refrain from putting ethics at the top of the work pile, or through cultures that heavily reward fast, unencumbered product delivery). In fact, in smaller companies moral AI work is often done on a volunteer basis, because there is no support for it to be completed as part of formal job descriptions.[19] Until the people who are creating, packaging, applying, scaling, and monitoring AI feel confident that the effort moral AI requires is consistent with what their organizations want them to prioritize, ethical issues are likely to go largely unaddressed, regardless of what technical tools are available or what guiding principles are advertised.

Lean-agile methods often exacerbate this problem. When

plans to address ethical concerns do not facilitate new product iterations or do not improve other established metrics at the speed at which lean-agile product iterations are typically expected in these environments, those ethical concerns are likely to end up being viewed as impediments to progress and threats to job security rather than inspiration for innovation.[20] We suspect this is a major reason many top AI companies ultimately reduce or even eliminate their AI ethics teams, despite adopting ethical AI principles at a broad policy level.[21]

POSTPONING ETHICS

Even within organizations that genuinely want to prioritize ethics, AI product teams are very motivated to choose paths that allow them to determine quickly whether their AI project has the potential to succeed. Figuring out how to make an *ethical* AI model currently takes more time and resources than figuring out how to make *any* functioning AI model, so many teams will elect (or be forced) to postpone thinking about ethical issues until after they succeed in creating the most straightforward, sufficiently accurate AI model.[22]

In principle, many ethical considerations can still be appropriately managed in later stages of product development, but there are two dangers to postponing ethical concerns to later stages. First, responding to ethical issues might never be considered a necessary step in creating the current minimum viable product iteration. This phenomenon contributes to a second danger: lock-in effects. Once a 'working' model is achieved and is receiving positive feedback, it can feel prohibitively costly to change the model or product. Internal stakeholders then become highly motivated to make easy changes to established 'unethical' models instead of incurring the costs of developing new product versions that address ethical issues more adequately.

NO METRICS FOR ETHICS

Remember how we said that any phenomenon that is not tracked by a metric is typically not given much attention by a product team? Well, at least at the time of writing this book, there were still no well-established metrics for teams or organizations to use to monitor the ethical impacts of their AI work processes or products, even if some teams have developed ad-hoc metrics to monitor very specific ethical issues, like the percentage of nude photos a content moderation system flags accurately. Without broader moral AI metrics that are supported and discussed throughout an organization, teams have no straightforward way to integrate moral AI concepts into their lean-agile processes and are less likely to have their financial compensation tied to ethical impacts of their work. More generally, organizations struggle to integrate ethical impacts of their products into their overall decision-making and team management if they do not have relevant objective measurements to draw on.

SEPARATING SOCIETAL AND PROFESSIONAL ETHICS

Why do we need moral AI metrics to do the right thing? Shouldn't we be able to be ethical without having our choices tracked so explicitly? Yes, but the ethics we have learned to pay attention to at work may not cover the broad range of ethical issues AI products raise. For example, some engineers or data scientists working on AI have been trained to believe the ethics of their profession do not include the greater social implications of individual projects they work on. In the words of one engineering professor:

> Our ethics have become mostly technical: how to design properly, how to not cut corners, how to serve our clients well. We work hard to prevent failure of the systems we build, but only in relation to what these systems are meant to do, rather than the way they might actually be utilized, or whether they should have been built at all. We are not amoral, far from it; it's just that we have steered ourselves into a place where our morality has a smaller scope.[23]

Similarly, AI designers sometimes consider their professional ethical responsibilities to be limited to commitments to their employer and their products' functionality, rather than including the concerns of society at large.[24] Even if engineers, data scientists, or designers do believe that their professional ethical responsibilities include making sure their products have socially acceptable consequences, few of them have had much experience identifying and analysing those consequences, so they go undetected or unaddressed.

INADEQUATE COMMUNICATION ABOUT ETHICAL INFORMATION

Now imagine that an organization and all of its AI contributors are truly committed to moral AI. Will the path to moral AI be smooth sailing? No.

In today's AI ecosystem, many different people with different backgrounds contribute to AI products in ways that build on previous contributions. Lean-agile processes also lead to team members working in parallel instead of in concert. Team communication in these contexts often looks like a gigantic game of telephone (known in some places as Chinese whispers) where the message unknowingly gets changed as it is passed from player to player. The result is that critical moral information (like a chosen fairness definition, collection bias within a data set, or unequal performance among demographic groups) frequently gets lost or misrepresented as a product gets passed among team members or decision-makers, making it hard to predict or fix moral problems.

AI algorithm developers, in particular, tend to have very little contact with those using or impacted by the AI algorithms they develop. When this is the case, it is very difficult for algorithm developers to learn enough about the nature of the algorithm's impacts to make modifications that can mitigate those impacts, so the technical tools they make available do not succeed at solving the problem they were intended to address or, worse, actively cause their own unique unethical consequences.

SHIRKING ETHICAL RESPONSIBILITY

In the game 'Nose Goes', somebody asks, 'Who wants to [fill in the blank with a task nobody wants]?' and everybody rushes to touch their nose and say 'Not it!' as quickly as they can, because the last person to do so has to complete the task.[25] Implementing moral AI can feel very similar, with everybody involved racing to say 'Not it!' first, even if everyone wants their AI products to be ethical in principle.[26] The motivation to avoid daily responsibility for implementing moral AI strategies is understandable, but crucial issues are unlikely to be addressed by anybody in the organization if AI contributors are not given pragmatic ways to assign responsibility.

One approach to this problem might be for the parent organization as a whole to assume legal and moral responsibility for its AI products' social impacts. Even if that approach is adopted, though, it does not clarify who within the organization will ensure the organization's moral AI principles are being built into its AI products on a daily basis. One of the challenges in assigning this responsibility is that respecting moral AI principles requires detailed technical decisions and knowledge at multiple steps along the product process. A CEO, board, or governmental office can make decisions about which general principles to follow, but it is difficult for them to wade as deep into the weeds of a product as these decisions require. It's great if a CEO commits to ensuring their organization's AI products are fair, but it seems unrealistic to expect a CEO to have enough technical background to understand the differences between twenty different mathematical definitions of fairness well enough that they can tell their engineering teams which one to use. Somehow, people who know the ins and outs of how a product is being made need to be in reliable communication with the people who will be held accountable for the moral outcomes of the product. It's not clear how to do that.

OK, then maybe technical teams should shoulder responsibility for the moral outcomes of an AI product. But who on the technical team should be responsible? The people who assemble the data may have little information about what the data will be used for. The people who decide how to organize and clean the data may have few details about where the data came from. The people who model the data may have incomplete information about the data they are given and little influence on how the model is used. The people who put the model into production may have little insight into the model itself or the data that was used to train it. This lack of knowledge about what others are doing makes it hard to figure out which team member is responsible when moral problems arise. Also, how should policies handle differences in the ways AI product teams interact – for example, that the people who train AI models put them into production and know where their data came from in one (part of an) organization but not in another?

Another issue is that although engineers and developers are tasked with building an AI product efficiently, it is the product manager's job to outline what the engineers should create in the first place.[27] So perhaps product managers should be responsible for at least some of the moral outcomes of an AI product. Unfortunately, however, they often lack the background or training to understand what kinds of technical issues can cause an AI to go awry (like biased data sets or biased model performance), and expecting all AI product managers to have a technical background is perhaps unrealistic.

Another approach would be to assign responsibility for different AI-related moral principles to different parts of an AI product team. For example, perhaps the engineering manager of the data and modelling teams is responsible for making sure an AI algorithm is fair, the product owner and user experience team are responsible for identifying adverse consequences the product could have on customers and society, and someone higher up in the organizational leadership is responsible for choosing how to address other moral issues the product team brings to their attention. This strategy seems reasonable but still faces challenges. Since each organization and product team is organized differently, each organization would have to figure out for every AI product it creates which parts of the team should take responsibility for which moral issues. Further, some important moral issues could get missed if groups create isolated parts of an AI system and then hand those parts over to other groups without communicating details that are necessary for the other groups to make good choices about the moral issues they are responsible for.

A different approach to assigning moral responsibility for an AI product is to say that everybody who contributes to the creation of the product must work together to ensure it has morally acceptable impacts. The danger with this approach is encompassed by the saying, 'If everybody is responsible, nobody is responsible.' Put simply, people are less likely to take responsibility for a task when it is not explicitly assigned to them, and they know they won't be held uniquely accountable. Moreover, it becomes more difficult for the actors to 'work together' to address ethical concerns when third parties are hired to perform individual tasks. When different organizations create different parts of an AI product, as is very common, each will have strong incentives to deny their own responsibility for an immoral outcome.

What do we need to bridge the moral AI principle-to-practice gap?

The above list of challenges is not comprehensive. It is only meant to convey the breadth of problems that moral AI technical tools and principles on their own do not address.[28] That raises the obvious question: what can we do? Thankfully, the short answer is: a lot! Here are five separate calls to action, each meant to inspire critical steps we need to take to make moral AI a reality.

Call to action 1: scale moral AI technical tools

Diverse technical ways to address AI's moral problems are increasingly available. Many of them have already been discussed in this book. These technical tools are not sufficient on their own to ensure that AI is used ethically. Nonetheless, extending their breadth, performance, and accessibility is essential for making moral AI scalable.

Some technical tools need to modify or leverage AI algorithms or models directly, like those that incorporate moral features into automated decisions, mathematically minimize unfairness, or facilitate model explanations. When these tools are made, they must be accompanied by 'translational' moral AI research that figures out what needs to happen in order for those tools to be implemented in the production-level settings of organizations, so that issues such as technical teams feeling unequipped to choose fairness definitions are brought to light and addressed.[29]

Other technical tools need to improve the development

and deployment processes that bring AI products to fruition. Examples of such process-focused tools include checklists that highlight the ethical issues most relevant to each part of the AI development process and that offer discussion topics to help AI teams identify ways those issues manifest in their particular applications;[30] explainable AI tools that enable AI teams and stakeholders to identify when AI models employ ethically questionable features like race, gender, or socio-economic status; 'data cards' that help teams summarize important decisions made about data sets which can impact the ethical use of AIs trained on them;[31] and auditing tools that determine whether an AI system is creating fair outcomes for different demographic groups.

Some worry that technical moral AI efforts will divert resources and attention away from grappling with systemic societal issues that underlie some of AI's negative impacts, or will incentivize 'ethical whitewashing' by empowering organizations to misleadingly claim their practices are ethical solely by virtue of using moral AI technical tools.[32] These fears are justified, and are why additional tactics must be built into any moral AI strategy. Nonetheless, these valid concerns shouldn't cause us to neglect the technical developments that are also required or helpful to realize moral AI.

At the same time, no matter how useful moral AI technical tools can be, they won't have a meaningful impact on the AI that we interact with on a daily basis if the organizations that create AIs don't have organizational practices that allow moral AI tools to be used effectively. That need is the focus of the next call to action.

Call to action 2: disseminate practices that empower moral AI implementation

This moral AI call to action involves widening knowledge of best organizational practices for generating moral AI products. Today's social and financial ecosystem typically prioritizes shareholders. This makes it challenging to recruit organizational leaders committed to moral AI. Thus, one goal of this call to action is to cultivate research into how organizational leaders committed to moral AI can most effectively be recruited and given a competitive advantage.[33] One helpful step in this regard would be to develop metrics and procedures for assessing moral AI performance that could be incorporated into CEOs' reviews and compensation contracts.

Another goal of this call to action is to build and disseminate evidence about what organizations that are genuinely committed to moral AI need to do to ensure their culture and operations lead to AI products that are ethical. To give a better idea of what we mean by this, here are some challenges such organizations face, and some tactics for overcoming them.

OPERATIONS ALIGNMENT

As we mentioned, AI contributors repeatedly report that moral AI principles are poorly aligned with their organizations' financial and business strategies, and with how AI creators are evaluated and compensated. A moral AI strategy should, therefore, create resources to help leaders: (i) assess discrepancies in their organizations' structures, practices, and ethical goals; and (ii) learn the most effective methods for convincing AI contributors that they are supported in allocating resources to anticipating and solving moral problems related to their AI products.

ORGANIZATIONAL CULTURE

Organizational leaders also need mentorship in fostering an environment for productive ethical deliberation when AI product issues do not have obviously 'right' answers.[34] This is tricky stuff, but some instructive insights have emerged.[35] Ethical deliberation is most successful when it engages contributors with different backgrounds and organizational roles,[36] yet asking diverse people to debate ethical issues can be counterproductive and divisive if not approached skilfully.[37] Trained facilitators are helpful to this end, but it is impractical to require their involvement in all ethical product decisions. Even when facilitators are available, participants have to be given resources before, during, and after the process to help them develop interpersonal skills supporting introspection, openness to feedback, and participation in sometimes emotionally charged disagreements.[38]

The success of such efforts is deeply affected by whether the participants feel safe and are empowered to take actions based on the results of the deliberations.[39] Establishing a culture of 'moral learning' also helps.[40] When people fear being labelled morally 'good' or 'bad', they will go to great lengths to avoid feedback and might even lie to maintain their moral standing. In contrast, when it is assumed that most people have good moral intentions but will make moral mistakes that can be corrected through practice, people are much more open to moral feedback and are much more likely to take action based on that feedback.

Moral AI strategies need to disseminate best practices based on insights like these, invest in research improving these best practices, and cultivate consulting resources. One beneficial move could be to commission university corporate relations teams to match organizations looking for moral AI guidance with academic researchers who can conduct impartial and rigorous research in organizational settings. Such collaborations would help organizations receive evidence-based guidance about how to meet their moral AI objectives, while also making it more likely the lessons learned will be widely shared, for example by publishing the research in relevant periodicals.

MAKING MORAL AI LEAN-AGILE COMPATIBLE

To remain viable, competitive, and motivated to change, AI-creating organizations need concrete, evidence-based guidance on how to adapt their established operational processes to achieve moral AI goals without severely handicapping their other organizational goals. One option in this regard is to develop an entirely new product management methodology that is as effective as lean-agile methodology and still incorporates ethical review and evaluation throughout a product's life cycle. A second option is to determine how to integrate moral AI into current lean-agile practices and culture. There is little systematic research on this, but here are some approaches that might facilitate this integration.

Hold all AI contributors responsible and organizational leaders accountable. We think that moral AI should be promoted as a shared responsibility of everybody who is involved in an AI product, but that organizational leaders should agree to be publicly accountable for an AI product's bad moral outcomes. Holding leaders accountable will relieve individual contributors' fears that they will be scapegoated when mistakes are inevitably made. Simultaneously, distributing moral AI responsibilities across an organization will alleviate the 'Not it!' problem and allow leaders to require each part of their organization to have a tailored plan for learning about, monitoring, addressing, and innovating around moral concerns in their areas.

Such an approach will allow organizations to respond more nimbly and innovatively when ethical concerns arise and will help prevent exclusive 'ethics owners' from becoming product bottlenecks. Further, requiring ethical skill development and social impact plans from all AI contributors will enable organizations to evaluate and tie compensation to moral learning and plan compliance, which will broadcast the message that the organization takes social impacts seriously. At the same time, making organizational leaders accountable for ethical concerns will clarify whom outside entities should go to when ethical concerns are identified and will give leaders incentives for ensuring that each team's plan and organizational practices meet moral AI needs.

Make AI ethical by original design. Leaders should instruct their AI product teams to include ethical and social concerns in the initial design requirements of all AI products. When leaders identify these concerns as practical constraints that are of equal priority to engineering or financial constraints, prototypes will be less likely to qualify as 'minimum viable products' if they have strong potential for harming society.

Some AI contributors might still be concerned that addressing ethics at the outset makes it too difficult to get an AI project off the ground. However, these concerns are likely to be assuaged, and maybe even reversed, once contributors realize how much time is saved in the long run if they pay attention to ethical requirements from the outset of a project, and once they become more practised in knowing how to identify and address ethical issues early on.

Integrate ethical evaluation. Organizations committed to moral AI should integrate people with ethical decision-making background into product teams as much as possible. Many organizations appoint specialized ethics groups or boards to act as a combination of consultants, risk mitigators, and rubber stamps for the organizations' AI products once the products are developed enough to warrant oversight.

Having specialized 'ethics owners' like these makes sense in some contexts, but it also has strong disadvantages.[41] First, it suggests that ethics should be addressed outside the core AI product team. Second, it increases the likelihood that people making ethical evaluations will not be given all the technical details they need to make good decisions, because others will not know what information is important to share. Third, separating ethical concerns from the engineering teams and design teams eliminates the opportunity for the engineering and design teams to learn how to navigate ethical issues in their own work.

For these reasons, we agree with the growing consensus that the best way to create moral AI is to include at least one person with experience in thinking about ethical issues into a product team's week-to-week processes so they can stay abreast of the product's details and can help the product team address ethical concerns as early as possible in a product's life cycle.[42] If an organization does not have access to people with ethics training, it should invest in providing such training for people already on its product teams.

If an organization is not able to integrate people with ethics experience into product teams, these teams should continuously evaluate societal implications and solicit ethical consultations throughout the product life cycle, rather than waiting until the product is well under way or finished. Ideally, societal implications would be a standard part of each agile sprint review and sprint retrospective (the time at the end of a sprint when teams evaluate their progress and processes).

Track moral AI metrics. Once these changes are implemented, incorporating modifications of common lean-agile methods may help the rest of a moral AI approach fall into place organically. One such modification would be to require AI product owners to add metrics that index progress towards societal goals to the metrics they already track, strategize around, and are skilled in crafting. Since metrics have to be quantitative measurements, examples could include collecting scores on user quizzes about an AI product's privacy policy to make sure users understand the policy, analysing an AI's accuracy by demographic group to make sure it's acceptably accurate in all groups, or gathering ratings of users' ethical concerns about the product.

The metrics will inevitably be imperfect, but the process of collecting and tracking data relevant to the social impact of AI products will make it much more likely that an organization will identify negative social impacts before the impacts become too firmly established to modify. This identification is an essential step for turning moral concerns into issues that AI professionals will actually address on a daily basis. In addition, collecting relevant data regularly will allow teams to objectively evaluate different approaches, making it more likely they will find an iteration with a more positive social impact than the initial version.

Expand AI user experience researcher and designer roles. Another modification would be to charge user experience (UX) researchers and designers with using their unique skills to understand the full range of ways their AI products might interact with different groups in society, especially minority groups (given that AI teams are primarily populated by Caucasian men).[43] Brad Smith and Carol Ann Browne, respectively Microsoft's president and senior director of communications and external relations, shared the following major lesson from their organization's admitted ethical mistakes:

> We needed to get out and listen to what other people had to say, and do more to help solve the technology problems that needed to be solved. That meant building constructive working relationships with more people. But that was just the start. We had to understand perceptions and concerns. We needed to do a better job of solving small problems before they grew out of control. We sat down more frequently with governments and even our competitors to find common ground.[44]

Of all AI contributors, UX researchers and designers are best suited to manage this type of listening and understanding.[45] These professionals are already trained to listen empathically to users and to ask them questions that uncover their desires, concerns, pain points, and joy in relation to a product – a concept known as the 'user journey'. Our suggested modification of their roles would simply lead to them applying their skills to a greater variety of stakeholders and gathering information about a broader type of AI product 'journey'.

Understanding people's moral or social concerns is admittedly quite different from more typical user-interaction topics, such as whether users like the layout of a website landing page. So UX researchers and designers might need additional training. Further, AI product teams might need help innovating new approaches to soliciting adequately diverse and frequent feedback. Such modifications will take some trial and error to optimize, but, once implemented, they will empower AI product teams to respond to societal concerns quickly, efficiently, and without much disruption to the way they create their products.

Call to action 3: provide career-long training opportunities in moral systems thinking

The previous call to action would greatly benefit from AI contributors having experience wrestling with AI-related ethical issues and developing expertise in navigating them. Unfortunately, most AI contributors are not trained to identify and respond to the ethical implications of their work. In the words of Tracy Chou, an engineer, CEO, and founder:

So many of the builders of technology today are people like me; people haven't spent anywhere near enough time thinking about these larger questions of what it is that we are building, and what the implications are for the world . . . I now wish that . . . I'd had opportunities to debate my peers and develop informed opinions on philosophy and morality. And even more than all of that, I wish I'd even realized that these were worthwhile thoughts to fill my mind with – that all of my engineering work would be contextualized by such subjects.[46]

For AI contributors to be adept at empowering their technology to avoid moral harms to society, they need frequent opportunities to practise (and to receive feedback about) how they identify, navigate, and pursue or prevent the societal effects of AI products. One of the most effective ways to do this is to engage in 'systems thinking'.

Systems thinking is the art and science of designing solutions for complex, real-life settings where products interact with interconnected entities and interdependent incentives, constraints, and limitations. Systems thinking is often used to understand why products or policies fail, through paying particular attention to how impacts of the product feed off each other and cause unintended side effects over time, sometimes through impacting the larger environment the product is being used in. When applied to AI, systems thinking requires us to examine an AI's impacts 'from multiple perspectives, to expand the boundaries of our mental models, to consider the long-term consequences of our actions, including their environmental, cultural, and moral implications'.[47]

AI product teams need many skills to engage in systems thinking about ethics successfully. Most obviously, they need to be able to reliably identify moral issues that arise at different times and to analyse different people's possible responses to them. Teams also need to be able to anticipate and listen to stakeholders' unique perspectives and to incorporate others' viewpoints into their assessment of a problem and its solutions.[48] Once an ethically appropriate and practically feasible course of action is chosen, team members also need to use compelling communication and persuasion to secure buy-in for their chosen path forward. They also need to leverage emotional intelligence skills to bring the chosen action to fruition among conflicting social opinions and professional pressures.[49] This wide skill set has often been considered peripheral to the central work of engineers, data scientists, and AI specialists,[50] but it is essential to moral AI.

The goal of this call to action is to cultivate career-long educational opportunities for AI contributors to practise moral systems thinking. Many AI contributors are already in senior positions without having had such training, so it will be important to cultivate training pathways for multiple career stages. In addition, since this kind of training will be relatively new, insights from educational research should guide initial efforts, and funding should be allocated to study and share lessons about which strategies are most successful.

This research will be crucial because some attempts at moral skill training backfire. For example, ethical decision-making workshops administered at 21 institutions across the US made participants *more* likely to suggest deceiving, retaliating, and avoiding personal responsibility in response to moral issues.[51] Similarly, the further engineering students progressed in their ethics and engineering curricula at four well-known US universities, the *less* important they rated professional and ethical responsibilities, social consciousness, understanding of the consequences of technology, and understanding of how people use machines.[52] The leading hypothesis about why these training efforts were so counterproductive is that they focused on rules, regulations, or punishments related to ethical issues, instead of intrinsic motivations for 'doing the right thing'.[53] They also didn't address the social and emotional intelligence skills that need to be added to ethical judgment skills to yield ethical behaviour. Training in moral systems thinking will need to leverage research from the growing behavioural ethics field and be continuously evaluated and refined to avoid such outcomes.[54]

With those considerations in mind, we offer a few tactical suggestions for teaching systems thinking about moral AI. We share the enthusiasm various groups have expressed for integrating systems thinking concepts and experiences throughout AI-related technical courses, rather than restricting them solely to standalone courses or add-on modules.[55] Embedding social considerations throughout technical training helps AI creators internalize the expectation that optimizing social impact is a fundamental part of successful technical work and helps them develop habits for identifying and grappling with moral implications in their workflow. One approach to integration is to include selected ethical issues as design requirements in technical assignments and projects so that it becomes common for technical contributors to be evaluated on how their AI solution addresses moral issues. Another approach is to assign projects that task learners with developing technical solutions to moral issues and then defending them in front of panels of stakeholders and social science experts.

In non-academic settings, similar results can be achieved through learning experiences designed intentionally to involve diverse parts of an organization. A workshop could, for example, task designers, lawyers, business representatives, product managers, marketers, and engineers with explaining to each other their role in creating a specific AI product, sharing what excites and concerns them, and brainstorming together about how to make sure the product aligns with selected moral values. In all these activities, mentors must evaluate learners on how well they handle the ethical components of the activities just as rigorously as they evaluate them on technical components. Doing so will ensure that learners get enough feedback to improve. It will also communicate that ethics and systems thinking are as valued as technical prowess. Importantly, if learners are *not* evaluated rigorously on the moral aspects of their work, then training experiences are likely to be perceived as ethics 'whitewashing' – that is, as attempts to mislead others into thinking an organization genuinely prioritizes ethics when, in fact, it does not. If training in moral systems thinking is seen as insincere or a screen to hide ulterior motives, it will undermine the training and have counterproductive consequences.

Another place where AI contributors can learn systems thinking is on the job. For such learning to occur with the intended outcomes, though, projects must be pursued in a way that allows contributors to make mistakes, receive feedback, and be exposed to peers' opinions on the moral implications of what is being built. This type of exchange can be built into lean-agile methods by having all contributors meet at regular intervals during a product's creation process to share ideas and data about the social impacts of their product and to make sure the team is monitoring the appropriate metrics. Regular meetings allow non-technical contributors to learn about the types of decisions technical contributors make when engineering a product and ensure that technical contributors repeatedly hear users' reactions to what they are building. Over time, the team as a whole learns which parts of an AI product life cycle are most likely to raise moral issues and which values are most important to them, personally and collectively.

To create enough training opportunities in moral AI systems thinking, we need a big pool of interdisciplinary instructors or facilitators with the appropriate experience and skill sets. Very few people currently have been trained to have all the technical knowledge and systems thinking skills we have discussed. As a result, generating this pool will be a challenge that will require deliberate financial, intellectual, and other resource investment from governments, organizations, and educational institutions. One great way for organizations to do this in-house is to recruit chief learning officers with moral AI expertise (or give chief learning officers opportunities to gain additional graduate training in moral AI),[56] and then task them both with designing tailored training programmes in moral AI systems thinking and with training members of their organization to lead moral AI programmes for specific groups. Governments and academic institutions should also help cultivate AI systems thinking specialists by sponsoring prestigious fellowships or mid-career training opportunities for researchers in relevant technical and non-technical domains.

Call to action 4: engage civic participation throughout the AI life cycle

The previous calls to action focus on AI professionals in organizations and academia, but these are far from the only stakeholders. Given the pervasiveness of AI and the global connectedness of our daily lives, almost any world citizen has the potential to be affected by an AI product. This is true even when people don't use an offending AI product themselves, such as when lawyers are prevented from entering Madison Square Garden because facial recognition AI thinks they work for firms Madison Square Garden's owners are in litigation with, or when a logo artist loses their job to an AI artist.

These stakeholders' experiences and opinions are essential to any meaningful conversation about moral AI. As Mark Zuckerberg, the CEO of Meta, has said:

> What I would really like to do is find a way to get our policies set in a way that reflects the values of the community, so I am not the one making those decisions . . . I feel fundamentally uncomfortable sitting here in California in an office making content policy decisions for people around the world.[57]

Our challenge is that it is not yet clear how to solicit useful and reliable opinions from all of an AI product's diverse stakeholders. The fourth call to action for a moral AI strategy tries to establish some ways to do so. The overarching goal of this call to action is to make it both *easy* and *likely* for a wide variety of community members to share their opinions, fears, and hopes about specific AI applications at a rapid and reliable pace. The flow of information between AI creators and AI community stakeholders needs to be bidirectional, so it should be convenient both for AI creators to share their new AI ideas and prototypes and for stakeholders to give their feedback. Is such a goal feasible? Well, it won't be easy, but we believe it can (and must) be done.

Effective solutions will likely take advantage of online platforms. These allow individuals to engage from different locations at different times, permitting a greater range of people to participate. Further, infinite varieties of digital experiences can be created using online platforms, ranging from educational sessions and prototype demonstrations to virtual focus groups, town hall meetings, or surveys and feedback forums. Moreover, all of these can be kept confidential, if needed.

Some experiences should be designed to allow anybody in the relevant group to participate at any time. The online platforms SpeakUp, Austin, created by the city of Austin in Texas,[58] and Your Dublin, Your Voice, created by the city of Dublin in Ireland[59] are good examples. Community members simply go to the platform to see descriptions of projects or policies their city is considering and are encouraged to provide anonymous feedback about them. Other experiences should seek to engage underrepresented groups and mediate virtual group discussions, such as those offered by the Brussels-based CitizenLab,[60] an organization that 'provides cities and governments a digital participation platform to consult their citizens on local topics and include them in decision-making'.

So far, feedback is typically voluntary on these platforms, but civic participation can be scaled by helping platforms partner with the 'gig economy' to use financial compensation to entice deeper, wider, and faster feedback. Platforms like Amazon Mechanical Turk, Qualtrics, and Prolific already provide mechanisms for organizations to pay people with specific demographic characteristics for their product feedback or participation in online surveys. Moving forward, governments, organizations, and entrepreneurs should take advantage of and expand on these types of platforms to solicit fast feedback about AI policies or AI product prototypes. Following community-engaged research practices in other domains, AI-creating organizations could also partner with individual civic groups that represent specific communities or demographics and provide financial compensation in return for honest, reliable input from group members.[61]

To clarify, the primary purpose of AI civic participation is *not* necessarily to establish a consensus about how AI's impacts should be handled or adjudicated. It would be fortunate if such a consensus develops. However, given people's polarized views of AI, it is unrealistic to expect that such a consensus will always emerge in the timeframe in which AI products are being developed.

Instead, the primary goal of civic participation, as we are thinking about it here, is to identify as many moral and social effects of a given type of AI as possible while it is still being created and can be modified, ideally in timeframes that are compatible with lean-agile product development processes. Much of the work that needs to be done in this call to action is figuring out what kind and cadence of bidirectional information sharing is needed to identify the moral issues raised by AI products in time for AI teams to address them, as well as establishing platforms and best practices for supporting that information sharing. This is a relatively new area which holds potential for exciting interdisciplinary innovation. It is also an area that must be invested in to make moral AI sustainable.

Call to action 5: deploy agile public policy

The tactics we have discussed thus far will work best if everybody contributing to AI products has both good intentions and sufficient resources to pursue moral AI reliably. Since neither of those things will always be true, public policies that provide incentives, resources, guidelines, and regulations aimed at maximizing the positive impact of AI on human society are another essential part of a moral AI strategy.

Some critics who focus exclusively on the 'regulation' role of public policy fear that regulations will endanger AI's chances of making our lives better, by increasing costs and stifling innovation. Citing such fears, employees at Clarifai, an AI image recognition company, wrote in the *New York Times*:

> Regulation slows progress, and our species needs
> progress to survive the many threats that face us today . . .
> Technology regulation is notoriously ham-handed when
> it's made by people who don't understand it . . . We need

to be ethical enough to be trusted to make this technology on our own, and we owe it to the public to define our ethics clearly.[62]

Using this logic, some AI producers resist moral AI public policy outright.

We think that categorically resisting regulation is harmful not only for society but also for AI's overall progress. Governments generally want their constituencies to lead AI development, not stifle it. As one executive vice president of the US Chamber of Commerce wrote, 'Whoever leads in the advancement of AI will lead the global economy.'[63] Thus, governments generally will at least try to make regulations that promote their long-lasting AI leadership.

Well-calibrated regulations can support technical AI leadership in many ways. First, regulatory constraints can generate new creative and productive solutions that would otherwise have been avoided. Requirements and regulations motivate creators to think outside the box and try new ideas rather than simply follow the path of least resistance.[64]

Second, regulations allow companies to outsource part of the work needed to figure out how to make their AI products ethical. Companies don't have to dedicate as many resources to designing ethical guidelines to follow, since the regulations determine many of those guidelines for them. Moral AI guidance and regulation thereby free companies to focus their efforts more directly on their technology or business. Perhaps this is why many leaders, including the CEOs of Microsoft, Google, and IBM, have made public calls for AI regulation.[65] Even though large companies often get the most attention or have the loudest complaints, regulations also benefit the broader competitive landscape by levelling the playing field for small and early-stage companies who lack experience in and infrastructure for navigating ethical issues.

Well-calibrated regulations also support long-term AI adoption by engendering trust, which is critical for AI leadership. Private companies are under great pressure to prioritize financial stakeholder earnings, which sometimes conflicts with longer-term interests of society. It is unrealistic to assume that pursuing short-term financial gain will reliably lead to optimal moral outcomes for society. Government regulation will help assure AI users and affected communities that their interests will be protected when financial conflicts arise, making society more open to AI overall.

Of course, regulation is just one form of public policy. Governments can also provide guidelines about what constitutes ethical practice in AI, and develop safety, transparency, and privacy standards so that organizations don't have to do that work by themselves. In doing so, they can lay out feasible options for verifying how those standards can be met. Governments can also provide positive incentives by awarding contracts only to AI partners who demonstrate adherence to ethical principles, or by providing tax breaks for AI regulation compliance. In addition, governments can create infrastructure that makes moral practices easier for AI technology creators, such as by funding the development and dissemination of moral AI technical tools or making fair, vetted data sets in different domains publicly available. They can also subsidize moral training for people who work on AI systems, or business training for people who provide ethical evaluations of AI systems. These are just a few examples of ways in which public policies can support moral AI.

The biggest challenge facing AI policymaking is that public policies and laws take a long time to enact, and there are valid fears that incomplete or suboptimal policies will be difficult to correct. For moral AI policy to be effective, it needs to be nimble enough to keep pace with AI's rapid technology and industry development. Therefore, it is essential to incorporate 'agile policy' mechanisms that allow potential AI policies to be tested and improved in limited contexts before implementing them through more permanent mechanisms. Examples include the 'regulatory sandboxes' Singapore created to allow autonomous vehicles to be tested on their streets without changes to national laws, and adaptive regulations that manifest time-limited decisions through instruments like 'sunset clauses'.[66] A different approach would be to sponsor experimental regulatory markets where private corporations compete to provide high-quality regulatory AI services.[67] If implemented, such an approach would allow governments to focus on regulating the regulators instead of the entire AI industry, and AI regulation would benefit from the speed of private innovation.

Agile public policy will not perfectly reflect moral principles, but society will make faster progress towards moral AI if we have imperfect AI public policies in place than if we have no moral AI policy incentives or regulations at all.[68] It is also important to acknowledge that organizations, policymakers, and citizens lose patience with moral AI policy initiatives when they do not yield results or they seem completely detached from contemporary AI uses, and agile policy mechanisms make it less likely that this will happen. In sum, we should not let the best become the enemy of the good.

The big picture

Whether AI creators realize it or not, they cannot make AI systems without making some decisions or assumptions about what is morally right or wrong. Further, all AI systems will inevitably have morally relevant effects. The idea behind moral AI is that decisions about AI systems that have moral consequences should be made intentionally and thoughtfully, rather than by accident or default. Moral AI is neither solely a technical problem nor solely an ethical one. It is a complex challenge which intersects with many disciplines, societal dynamics, and aspects of human nature. To make AI consistent with society's values, we need to incorporate systems-level approaches that take the interaction between different stakeholders and an AI's social context into account. The five calls to action we have described in this chapter provide essential mechanisms for doing that successfully.

It's up to us

We started this book saying AI deserves both optimism and pessimism. We have deep enthusiasm for all the good AI has the potential to do and, simultaneously, serious concerns about the damage AI can cause to our safety, privacy, and justice, and about the ability to hold people accountable for AI's damage. We have tried to explain those concerns and outline a strategy for at least partially addressing them. As any soldier will tell you, though, even the best reconnaissance and battle plan aren't enough to win a battle. You also need the right mindset to help you navigate individual decisions. To illuminate what that mindset should be, we can look to history. In the past, we have often encountered transformational technologies that brought both benefits and burdens.

Consider the late eighteenth and early nineteenth centuries. Revolutionary machinery like the spinning jenny inspired the creation of mechanical textile mills, which dramatically increased the availability of textiles across the world. As a result, people from all economic backgrounds could afford more clothing and spend less time growing and generating their own materials to make it. In addition, textile mills created more consistent jobs than agricultural work, drew people to urban areas where they had better access to education and medicine, and created new opportunities for unprecedented economic growth. All of this was tremendously exciting and was part of the 'machine age' that led to better wages across society, increased prosperity and economic growth, and unprecedented innovation. But the transformation wasn't all good. The 'machine age' resulted in terrible pollution, tragic working and living conditions, exploitative child labour, and increasingly unequal distributions of income and social power. Few previous times in history were more transformational, for better and for worse.

Also consider the 'better living through chemistry' era during the first half of the twentieth century. New understanding of the periodic table led to society-altering chemical technology like leaded gasoline, freon-powered refrigeration, and aerosols. Suddenly, automobiles became attractive and accessible to the average family, food could be kept longer and more safely, people finally received access to medical treatments requiring refrigeration or aerosol delivery, and shiny new consumer products – like hairspray – filled the shelves. Innovation and economic prosperity boomed. Unfortunately, so did lead poisoning, a side effect of leaded gasoline known even at the time. So, too, did the destruction of our planet's atmosphere. Unlike lead poisoning, this side effect was unanticipated, at least at first. Concentrations of freon and other chlorofluorocarbons in the Earth's atmosphere were increasing by many folds due to the wide adoption of the era's new chemical technology products. Those increased concentrations ultimately caused holes in our Earth's ozone layer, which is needed to protect us from ultraviolet light from the sun, and contributed to global warming.

These times in history teach lessons that apply to our current age of AI. First, some of a technology's most significant impacts can be known in advance, but others cannot. Second, even in retrospect, with the benefit of decades of data proving all the lamentable side effects of the mechanical revolution and chemistry heyday, most people still think society was right to develop textile technology, effective and affordable automobile fuels, and methods of refrigeration. They think our quality of life and life expectancy are significantly better because of these developments, despite their drawbacks. The problems were with *how* we pursued the technologies, not with *whether* we pursued the technologies at all.

Our view about AI should be similar. Carefully developed AI technology should be nourished and heavily invested in so that it can help usher in new and exciting waves of societal advancement. Banning all forms of AI to protect us from its dangers would be not only infeasible at this point, but foolish and possibly even immoral.

At the same time, we have now seen that individual technologies can have mindbogglingly far-reaching consequences. It is easy to fool ourselves into thinking that no one type of technology can have so much influence and that we don't actually bear responsibility for foreseeing harms of the future because they are too hard to predict accurately. That self-deception is particularly tempting when the immense consequences of technologies like AI are not yet fully apparent.

We also face another formidable challenge: our own human weaknesses. We are not very reliable at making good ethical decisions that maximize good for all when our own financial or social gains – or even the mere possibility of them – are at stake. We are also not very good at forgoing immediate rewards for larger future ones, particularly when a significant proportion of the latter benefit other people. There are tremendous opportunities to make a lot of money from AI very quickly, even while others are hurt in the process. This situation does not set up our human moral judgment faculties for success.

Finally, there is an elephant in the room that doesn't get talked about enough. At least for a lot of people, developing AI is really fun to be a part of. It requires wrestling with stimulating intellectual and scientific questions, usually with other people who are clever and fascinating in their own right. The pursuit feels like an exciting adventure, even if you are just watching it from the sidelines. J. Robert Oppenheimer, who led work on the atomic bomb, once said, 'When you see something that is technically sweet, you go ahead and do it and you argue about what to do about it only after you have had your technical success.'[1] Historian Steven Shapin concluded about the development of nuclear weapons, 'It was that fun – that total absorption in the elaborately funded "technically sweet" – which kept potential moral reflectiveness in check.'[2] The same thing is happening now with AI. Indeed, this eclipsing of moral concerns by technological sweetness is even more concerning today, because it is not only a few nuclear scientists but most of society that is tempted to become so engrossed by AI progress that they forgo addressing AI's potential abuses and harms.

Put together, there are a lot of incentives either to ignore the reasons to be pessimistic about AI or to tell ourselves that AI's possible negative impacts are a foregone conclusion that we can't do anything about. We must not accept that line of thinking. We can build AI that can morally self-regulate by avoiding actions it predicts we would see as immoral, but it will take patience and dedicated research. We can also design regulations, organizational practices, educational resources, and democratic technologies that will make it much less likely that AI will be used in harmful and immoral ways. To help us do those things in the best way possible, we can also deepen our scientific understanding of how AI systems work and the ways in which they can be directed towards human benefits rather than harms. These, too, are tremendously exciting scientific pursuits that are fun to engage in without having to scale ever larger and more powerful AI systems. They may even help us gain insights into age-old mysteries about human intelligence and consciousness.

Yes, these goals will take discipline and commitment to realize, and need to be coordinated, but we think the investment is clearly worth it. These steps will not be sufficient to prevent all immoral AI, but they will get us a long way towards ensuring AI's net impact on society is sufficiently positive that future generations will look back and agree we are on the right side of history.

That brings us to the mindset we need to have as we engage in these pursuits. It may seem like this book is about AI. In many ways, of course, it is. But the AIs of today cannot interact with or affect society without our help. Humans still have to build AI models, train AI models, power AI models, and make AI models accessible to others for AI to do much of anything. Even if an AI of the future does eventually take over the world, we will first have to create that AI. So how the story unfolds will depend just as much on human moral decision-making and moral intelligence as it does on artificial decision-making and artificial intelligence. We should be clear about our role when we think about AI. In the end, humans are both the directors and the protagonists of the moral AI story. AI is just along for the ride – for now.

References

INTRODUCTION: WHAT'S THE PROBLEM?

1. See video at: https://www.youtube.com/watch?v=X_d3MCkIvg8
2. Klein, Alice. 'Tesla driver dies in first autonomous car crash in US.' *New Scientist*, 1 July 2016.
3. http://www.bbc.com/future/story/20160714-what-does-a-bomb-disposal-robot-actually-do
4. Schachtman, Noah. 'Robot cannon kills 9, wounds 14.' *Wired*, 18 October 2007.
5. https://sites.duke.edu/quantifyinggerrymandering/
6. https://fortune.com/2018/04/10/facebook-cambridge-analytica-what-happened/; Hu, Margaret. 'Cambridge Analytica's black box.' *Big Data and Society* (2020): 1–6. It is not clear whether this tool had any real effect on the election, but this case raises fears that AI could manipulate voters. Of course, Donald Trump's supporters might not see his election victory as bad news, but they could still worry when their political opponents use similar tactics.
7. https://www.thoughtriver.com
8. *Loomis v. Wisconsin*, 881 N.W.2d 749 (Wis. 2016), cert. denied, 137 S.Ct. 2290 (2017). We will discuss this case in more detail in Chapter 4.
9. https://www.asianscientist.com/2016/08/topnews/ibm-watson-rare-leukemia-university-tokyo-artificial-intelligence/
10. Obermeyer, Ziad, et al. 'Dissecting racial bias in an algorithm used to manage the health of populations.' *Science* 366(6464) (2019): 447–453.
11. Thanks to Coleman Kraemer for this example.
12. https://en.wikipedia.org/wiki/2010_Flash_Crash
13. https://www.cais.usc.edu/projects/hiv-prevention-homeless-youth/
14. Hill, Kashmir. 'How target figured out a teen girl was pregnant before her father did.' *Forbes*. https://www.forbes.com/sites/kashmirhill/2012/02/16/

how-target-figured-out-a-teen-girl-was-pregnant-before-her-father-did/#1d3ac9136668. Thanks to Jillian Kohn for this example.

15. https://www.cnn.com/2020/12/24/entertainment/spotify-ai-bot-judges-your-taste-in-music-trnd/index.html

16. https://www.christies.com/features/a-collaboration-between-two-artists-one-human-one-a-machine-9332-1.aspx

17. https://www.latimes.com/la-me-quakebot-faq-20190517-story.html

18. https://www.wired.com/story/deepfakes-cheapfakes-and-twitter-censorship-mar-turkeys-elections/

19. Cormier, Zoe. 'The technology fighting poachers.' bbcearth.com

20. Mozur, Paul. 'One month, 500,000 face scans: how China is using A.I. to profile a minority.' *New York Times*, 14 April 2019

21. In response, some governments are considering making it illegal to collect facial image data, e.g. Illinois in the US (https://www.ilga.gov/legislation/ilcs/ilcs3.asp?ActID=3004&ChapterID=57). In the EU, the proposed AI Act, Title II, Article 5, prohibits 'the use of 'real-time' remote biometric identification systems in publicly accessible spaces for the purpose of law enforcement' except in special conditions (https://artificialintelligenceact.eu/the-act/).

22. Gagliordi, Natalie. 'How self-driving tractors, AI, and precision agriculture will save us from the impending food crisis.' *Tech Republic*, December 2018.

23. Hao, By Karen. *MIT Technology Review*, 6 June 2019.

CHAPTER 1: WHAT IS AI?

1. http://www-formal.stanford.edu/jmc/history/dartmouth/dartmouth.html

2. https://www.congress.gov/bill/116th-congress/house-bill/6216/text

3. https://www.washingtonpost.com/technology/2022/06/11/google-ai-lamda-blake-lemoine/

4. Korf, R. E. 'Does Deep Blue use artificial intelligence?' *ICGA Journal* 20(4) (1997): 243–5.

5. e.g. https://ieeexplore.ieee.org/stamp/stamp.jsp?arnumber=8340798

6. https://herbertlui.net/9-examples-of-writing-with-openais-gpt-3-language-model/

7. Marcus, Gary, and Ernest Davis. 'Experiments testing GPT-3's ability at commonsense reasoning: results' (2020).

8. Reed, Scott, et al. 'A generalist agent.' arXiv preprint arXiv:2205.06175 (2022).

9. https://www.zdnet.com/article/deepminds-gato-is-mediocre-so-why-did-they-build-it/

10. https://quoteinvestigator.com/2011/11/05/computers-useless/; https://www.quora.com/Pablo-Picasso-stated-Computers-are-useless-They-can-only-give-you-answers-Is-this-a-valid-judgement

11. Starnino, Carmine. 'Robots are writing poetry and many people can't tell the difference.' *The Walrus*, 5 May 2022. https://thewalrus.ca/ai-poetry/

12. Chen, B., C. Vondrick, and H. Lipson. 'Visual behavior modeling for robotic theory of mind.' *Scientific Reports* 11, (1) (2021), 424; https://www.discovermagazine.com/technology/have-ai-language-models-achieved-theory-of-mind

13. https://arxiv.org/pdf/1602.04938.pdf

14. https://www.newscientist.com/article/2111041-glasses-make-face-recognition-tech-think-youre-milla-jovovich/; Sharif, Mahmood, et al. 'Accessorize to a crime: real and stealthy attacks on state-of-the-art face recognition.' In *Proceedings of the 2016 ACM SIGSAC Conference on Computer and Communications Security* (2016): 1528–40.

15. Like Christopher Manning, a Stanford computer science professor. See Mitchell, M., and D. C. Krakauer. 'The debate over understanding in AI's large language models.' *Proceedings of the National Academy of Sciences* 120(13) (2023), p.e2215907120.

16. Michael, J., et al. 'What do NLP researchers believe? Results of the NLP community metasurvey.' arXiv preprint arXiv:2208.12852 (2022).

CHAPTER 2: CAN AI BE SAFE?

1. Bostrom, N. *Superintelligence: Paths, Dangers, Strategies*. Oxford University Press, 2014.

2. Ovid. *Metamorphoses, Book* XI.

3. https://futureoflife.org/2016/12/12/artificial-intelligence-king-midas-problem/

4. https://www.wired.com/brandlab/2015/05/andrew-ng-deep-learning-mandate-humans-not-just-machines/

5. https://news.bloombergtax.com/tax-insights-and-commentary/we-can-all-learn-a-thing-or-two-from-the-dutch-ai-tax-scandal

6. Attewell, Paul. 'The deskilling controversy.' *Work and Occupations* 14(3) (1987): 323–46. See also Vallor, Shannon. *Technology and the Virtues: A*

Philosophical Guide to a Future Worth Wanting. Oxford University Press, 2018.

7. Dahmani, L., and V. D. Bohbot. 'Habitual use of GPS negatively impacts spatial memory during self-guided navigation.' *Scientific Reports* 10(1) (2020): 1–14; Sugimoto, M., et al. 'Online mobile map effect: how smartphone map use impairs spatial memory.' *Spatial Cognition and Computation* 22(1–2) (2022): 161–83.

8. https://www.trendmicro.com/vinfo/us/security/news/cybercrime-and-digital-threats/exploiting-ai-how-cybercriminals-misuse-abuse-ai-and-ml; https://www.darkreading.com/analytics/passgan-password-cracking-using-machine-learning; https://inews.co.uk/news/thermal-attack-technology-scammers-crack-passwords-pins-heat-fingers-warn-experts-1903250

9. https://www.bbc.com/news/business-64464140

10. https://www.wsj.com/articles/BL-MB-21942

11. Huang, L., Z. Lu and P. Rajagopal. 'Numbers, not lives: AI dehumanization undermines covid-19 preventive intentions.' *Journal of the Association for Consumer Research* 7(1) (2022): 63–71.

12. https://www.who.int/news-room/fact-sheets/detail/road-traffic-injuries

13. https://news.un.org/en/story/2021/10/1102522

14. https://www.linkedin.com/pulse/how-ai-could-have-saved-boeing-737-max-chad-steelberg

15. Ebbatson, M., et al. 'The relationship between manual handling performance and recent flying experience in air transport pilots.' *Ergonomics* 53(2) (2010): 268–77.

16. Ebbatson, M. *The Loss of Manual Flying Skills in Pilots of Highly Automated Airliners.* PhD thesis, Cranfield University, 2009.

17. Oliver, Nick, Thomas Calvard and Kristina Potočnik. 'The tragic crash of Flight AF447 shows the unlikely but catastrophic consequences of automation.' *Harvard Business Review,* 15 September 2017.

18. 'Automated flying creating new errors, NTSB chief says.' CBS News, 24 June 2014.

19. Randazzo, Ryan. 'What went wrong with Uber's Volvo in fatal crash? Experts shocked by technology failure.' *Arizona Republic,* 23 March 2018. https://eu.azcentral.com/story/money/business/tech/2018/03/22/what-went-wrong-uber-volvo-fatal-crash-tempe-technology-failure/446407002/; https://www.ntsb.gov/investigations/accidentreports/reports/har1903.pdf

20. Metz, Cade. 'The costly pursuit of self-driving cars continues on. And on. And on.' *New York Times*, 24 May 2021.

21. Wilson, B., J. Hoffman and J. Morgenstern. 'Predictive inequity in object detection.' arXiv preprint arXiv:1902.11097 (2019).

22. Gates, D. 'Flawed analysis, failed oversight: How Boeing, FAA certified the suspect 737 MAX flight control system.' *Seattle Times*, 21 March 2019.

23. https://docs.google.com/document/d/1yXni1GoD93q8mX-yom7JLBno Q8tPOQz2A_y3m3LJi8o/edit; https://electrek.co/2020/08/27/tesla-hack-control-over-entire-fleet/

24. https://keenlab.tencent.com/en/whitepapers/Experimental_Security_Research_of_Tesla_Autopilot.pdf

25. https://warisboring.com/this-new-backpack-robot-can-clear-minefields/

26. https://www.popsci.com/story/technology/mq-25-stingray-set-to-launch-2024/

27. https://www.cna.org/reports/2022/02/leveraging-ai-to-mitigate-civilian-harm

28. https://www.theguardian.com/uk/2006/oct/31/military.iraq

29. https://www.upi.com/Defense-News/2003/04/24/Feature-The-Patriots-fratricide-record/63991051224638/

30. https://www.cnas.org/publications/reports/patriot-wars

31. https://foreignpolicy.com/2018/03/28/patriot-missiles-are-made-in-america-and-fail-everywhere/; https://www.nbcnews.com/think/opinion/trump-sending-troops-saudi-arabia-shows-short-range-air-defenses-ncna1057461

32. https://dsb.cto.mil/reports/2000s/ADA435837.pdf

33. https://www.cnas.org/publications/reports/patriot-wars

34. https://www.cnas.org/publications/reports/patriot-wars

35. Of note, the US military at least initially refused to donate the Patriot system, specifically, to Ukraine, citing that they wouldn't have enough time to learn how to use it: https://www.defenseone.com/threats/2022/03/why-us-wont-give-patriot-interceptors-ukraine/363042/

36. N2103772.pdf (un.org)

37. https://www.forbes.com/sites/davidhambling/2021/07/21/israels-combat-proven-drone-swarm-is-more-than-just-a-drone-swarm/?sh=7eae470a1425

38. Khan, Azmat. 'Hidden Pentagon records reveal patterns of failure in deadly airstrikes.' *New York Times*, 18 December 2021.

39. Gould, L., and N. Stel. 'Strategic ignorance and the legitimation of remote warfare: the Hawija bombardments.' *Security Dialogue* 53(1) (2021): 57–74; Khan, Azmat. 'Hidden Pentagon records reveal patterns of failure in deadly airstrikes.' *New York Times*, 18 December 2021.

40. Khan, Azmat. 'Hidden Pentagon records reveal patterns of failure in deadly airstrikes.' *New York Times*, 18 December 2021.

41. https://nationalinterest.org/blog/reboot/why-us-air-force-pilot-intentionally-fired-patriot-missile-battery-198074

42. Philipps, Dave. 'The unseen scars of the remote-controlled kill.' *New York Times*, 17 April 2022.

43. https://www.washingtonpost.com/technology/2021/07/07/ai-weapons-us-military/

44. Morrison, E. E. *Health Care Ethics: Critical Issues for the 21st Century.* Jones & Bartlett Learning, 2009.

45. https://www.whatnextglobal.com/post/modern-ai-enabled-pacemakers

46. https://news.harvard.edu/gazette/story/2020/11/risks-and-benefits-of-an-ai-revolution-in-medicine/

47. Caruana, R., et al. 'Intelligible models for healthcare: predicting pneumonia risk and hospital 30-day readmission.' *Proceedings of the 21th ACM SIGKDD International Conference on Knowledge Discovery and Data Mining* (2015): 1721–30.

48. Gaube, S., et al. 'Do as AI say: susceptibility in deployment of clinical decision-aids.' *NPJ Digital Medicine* 4 (2021): 31.

49. Tobia, K., A. Nielsen, and A. Stremitzer. 'When does physician use of AI increase liability?' *Journal of Nuclear Medicine* 62(1) (2021): 17–21.

50. Povyakalo, A. A., et al. 'How to discriminate between computer-aided and computer-hindered decisions: a case study in mammography.' *Medical Decision Making* 33(1) (2013): 98–107.

51. Tsai, T. L., D. B. Fridsma, and G. Gatti. 'Computer decision support as a source of interpretation error: the case of electrocardiograms.' *Journal of the American Medical Informatics Association* 10(5) (2003): 478–83.

52. In the US, entities can use a fast-track process to market AI products with a moderate level of risk without ever testing the AI in clinical settings, as long as the AI is 'substantially equivalent' to other existing products or practices. https://www.fda.gov/medical-devices/premarket-submissions-selecting-and-preparing-correct-submission/premarket-notification-510k

53. https://www.scientificamerican.com/article/artificial-intelligence-is-rushing-into-patient-care-and-could-raise-risks/

54. https://medcitynews.com/2021/04/in-scramble-to-respond-to-covid-19-hospitals-turned-to-models-with-high-risk-of-bias/

55. Wynants, Laure, et al. 'Prediction models for diagnosis and prognosis of covid-19: systematic review and critical appraisal.' *BMJ* 369 (8242) (2020): m1382; Miller, J. L., et al. 'Prediction models for severe manifestations and mortality due to COVID-19: a systematic review.' *Academic Emergency Medicine* 29(2) (2022): 206–16.

56. Seyyed-Kalantari, L., et al. 'Underdiagnosis bias of artificial intelligence algorithms applied to chest radiographs in under-served patient populations.' *Nature Medicine* 27(12) (2021): 2176–82.

57. Vosoughi, S., D. Roy, and S. Aral. 'The spread of true and false news online.' *Science* 359(6380) (2018): 1146–51.

58. https://www.thenationalnews.com/world/asia/man-killed-in-india-s-latest-mob-attack-on-suspected-child-kidnappers-1.906836

59. https://www.cnn.com/2018/06/12/asia/india-whatsapp-facebook-false-kidnappings-intl-trnd/index.html

60. https://factly.in/unrelated-visuals-and-scripted-videos-are-being-shared-with-false-child-kidnapping-rumours/

61. https://www.france24.com/en/20180714-death-fake-news-social-media-fuelled-lynchings-shock-india

62. Gisondi, M. A., et al. 'A deadly infodemic: social media and the power of COVID-19 misinformation.' *Journal of Medical Internet Research* 24(2) (2022): e35552.

63. Islam, M. S., et al. 'COVID-19–related infodemic and its impact on public health: a global social media analysis.' *American Journal of Tropical Medicine and Hygiene* 103(4) (2020): 1621–9.

64. Jiang, R., et al. 'Degenerate feedback loops in recommender systems.' In *AIES '19: Proceedings of the 2019 AAAI/ACM Conference on AI, Ethics, and Society* (2019): 383–90; Cinelli, Matteo, et al. 'The echo chamber effect on social media.' *Proceedings of the National Academy of Sciences* 118(9) (2021): e2023301118; Hao, Karen. 'Deep Mind is asking how AI helped turn the internet into an echo chamber.' *MIT Technology Review*, 7 March 2019.

65. Rathje, Steve, Jay J. Van Bavel, and Sander van der Linden. 'Out-group animosity drives engagement on social media.' *Proceedings of the National Academy of Sciences* 118(26) (2021): e2024292118.

66. Atari, M., et al. 'Morally homogeneous networks and radicalism.' *Social Psychological and Personality Science* 13(6) (2021): 999–1009.

67. Awan, I., and I. Zempi. '"I will blow your face OFF": VIRTUAL and physical world anti-Muslim hate crime.' *The British Journal of Criminology*

57(2) (2017): 362–80; Williams, M. L., et al. 'Hate in the machine: anti-black and anti-Muslim social media posts as predictors of offline racially and religiously aggravated crime.' *British Journal of Criminology* 60(1) (2019): 93–117; Müller, K., and C. Schwarz. 'Fanning the flames of hate: social media and hate crime.' *Journal of the European Economic Association* 19(4) (2020): 2131–67.

68. https://www.newsweek.com/stephen-hawking-warns-artificial-intelligence-could-end-humanity-332082

CHAPTER 3: CAN AI RESPECT PRIVACY?

1. https://www.vice.com/en/article/kzm59x/deepnude-app-creates-fake-nudes-of-any-woman

2. https://contentmavericks.com/

3. https://graziadaily.co.uk/life/in-the-news/deepnudes-app/

4. https://www.merriam-webster.com/dictionary/privacy

5. Warren, Samuel D., and Louis D. Brandeis. 'The right to privacy.' *Harvard Law Review* 4(5) (1890): 195, citing Thomas M. Cooley, *A Treatise on the Law of Torts, or, The Wrongs Which Arise Independent of Contract* 2nd edn, Callaghan & Company, 1888, p. 29.

6. https://www.theatlantic.com/technology/archive/2013/02/why-does-privacy-matter-one-scholars-answer/273521/; Cohen, J. E. 'What privacy is for.' *Harvard Law Review* 126(7) (2013): 1904.

7. http://www.businessdictionary.com/definition/privacy.html

8. Reiman, Jeffrey H. 'Privacy, intimacy, and personhood.' *Philosophy and Public Affairs* 6(1) (1976): 26–44.

9. Nam, J. G., et al. 'Development and validation of deep learning-based automatic detection algorithm for malignant pulmonary nodules on chest radiographs.' *Radiology* 290(1) (2018): 218–28.

10. Wang, Yilun, and Michal Kosinski. 'Deep neural networks are more accurate than humans at detecting sexual orientation from facial images.' *Journal of Personality and Social Psychology* 114(2) (2018): 246.

11. https://www.facebook.com/help/122175507864081; https://www.npr.org/sections/thetwo-way/2017/12/19/571954455/facebook-expands-use-of-facial-recognition-to-id-users-in-photos; https://www.vox.com/future-perfect/2019/9/4/20849307/facebook-facial-recognition-privacy-zuckerberg

12. https://support.apple.com/en-us/HT208109

13. https://www.cnn.com/travel/article/airports-facial-recognition/index.html

14. https://theintercept.com/2018/03/06/new-orleans-surveillance-cameras-nopd-police/; https://www.nbcnews.com/news/us-news/how-facial-recognition-became-routine-policing-tool-america-n1004251; https://www.wired.com/story/some-us-cities-moving-real-time-facial-surveillance/; https://www.baltimoresun.com/news/crime/bs-md-facial-recognition-20161017-story.html

15. https://www.facefirst.com/industry/face-recognition-for-casinos/; https://www.reviewjournal.com/business/casinos-gaming/facial-recognition-technology-coming-to-las-vegas-strip-casinos/; https://calvinayre.com/2019/07/18/casino/macau-casinos-to-implement-facial-recognition-software/

16. https://thelensnola.org/2018/10/24/months-after-end-of-predictive-policing-contract-cantrell-administration-works-on-new-tool-to-id-high-risk-residents/. New Orleans is not alone. See also Tampa Bay: https://projects.tampabay.com/projects/2020/investigations/police-pasco-sheriff-targeted/intelligence-led-policing/

17. https://www.nola.com/news/crime_police/article_c39369dd-b5da-5322-bd33-da2e75b1f435.html

18. https://www.theverge.com/2016/10/11/13243890/facebook-twitter-instagram-police-surveillance-geofeedia-api; https://www.nytimes.com/2016/10/12/technology/aclu-facebook-twitter-instagram-geofeedia.html; https://www.aclunc.org/blog/facebook-instagram-and-twitter-provided-data-access-surveillance-product-marketed-target; https://www.worldwatchmonitor.org/2020/01/risk-of-persecution-going-digital-with-rise-of-surveillance-state/

19. The main arguments we will offer supporting this conclusion are summarized from the analysis detailed in Hirose, Mariko. 'Privacy in public spaces: the reasonable expectation of privacy against the dragnet use of facial recognition technology.' *Connecticut Law Review* 49(5) (2017): 1591–1620. https://core.ac.uk/download/pdf/302394726.pdf

20. *Katz v. United States*, 389 U.S. 347 (1967). Citation retrieved from Hirose, Mariko. 'Privacy in public spaces: the reasonable expectation of privacy against the dragnet use of facial recognition technology.' *Connecticut Law Review* 49(5) (2017): 1591.

21. *Smith v. Maryland*, 442 U.S. 735, 740 (1979) (internal quotation marks omitted) (quoting Katz, 389 U.S. at 361 [Harlan, J., concurring]). Citation

retrieved from Hirose, Mariko. 'Privacy in public spaces: the reasonable expectation of privacy against the dragnet use of facial recognition technology.' *Connecticut Law Review* 49(5) (2017): 1602.

22. *United States v. Maynard*, 615 F.3d 544, 555–56 (D.C. Cir. 2010), aff'd on other grounds sub nom. *United States v. Jones*, 132 S. Ct. 945 (2012). Citation retrieved from Hirose, Mariko. 'Privacy in public spaces: the reasonable expectation of privacy against the dragnet use of facial recognition technology.' *Connecticut Law Review* 49(5) (2017): 1605.

23. *United States v. Maynard*, 615 F.3d 544, 555–56 (D.C. Cir. 2010), aff'd on other grounds sub nom. *United States v. Jones*, 132 S. Ct. 945 (2012). Citation retrieved from Hirose, Mariko. 'Privacy in public spaces: the reasonable expectation of privacy against the dragnet use of facial recognition technology.' *Connecticut Law Review* 49(5) (2017): 1605.

24. https://www.mircomusolesi.org/papers/ubicomp18_autoencoders.pdf

25. *United States v. Maynard*, 615 F.3d 544, 555–56 (D.C. Cir. 2010), aff'd on other grounds sub nom. *United States v. Jones*, 132 S. Ct. 945 (2012). Citation retrieved from Hirose, Mariko. 'Privacy in public spaces: the reasonable expectation of privacy against the dragnet use of facial recognition technology.' *Connecticut Law Review* 49(5) (2017): 1605.

26. Newman, Lily Hay. 'AI wrote better phishing emails than humans in a recent test.' *Wired*, 7 August 2021. https://www.wired.com/story/ai-phishing-emails/

27. Fredrikson, M., S. Jha, and T. Ristenpart. 'Model inversion attacks that exploit confidence information and basic countermeasures.' In *Proceedings of the 22nd ACM SIGSAC Conference on Computer and Communications Security* (2015): 1322–33.

28. The public data set was collected by FiveThirtyEight. The AI model that was subjected to a model inversion attack was a decision tree that BigML makes available to its customers. BigML is an AI-as-a-service (AIaaS) platform that allows users to upload their data sets, train an AI model to predict selected variables, and then make the resulting tree available for others to use, all without users having to know how to design and train AIs themselves. It is important to emphasize that a model inversion attack couldn't be used on its own to successfully determine whether you cheated on your significant other unless your cheating data were used to train the AI model being attacked.

29. Brown, G., et al. 'When is memorization of irrelevant training data necessary for high-accuracy learning?' In *Proceedings of the 53rd Annual ACM SIGACT Symposium on Theory of Computing* (2021): 123–32.

30. https://www.businesswire.com/news/home/20130515006369/en/Nuix-and-EDRM-Republish-Enron-Data-Set-Cleansed-of-More-Than-10000-Items-Containing-Private-Health-and-Financial-Information

31. https://www.businessinsider.com/chatgpt-microsoft-warns-employees-not-to-share-sensitive-data-openai-2023-1

32. For examples of the momentum in this space, see De Cristofaro, Emiliano. 'A critical overview of privacy in machine learning.' *IEEE Security and Privacy* 19(4) (2021): 19–27; or Ma, Chuan, et al. 'Trusted AI in multi-agent systems: an overview of privacy and security for distributed learning.' arXiv preprint arXiv:2202.09027 (2022).

33. https://thereboot.com/why-we-should-end-the-data-economy/

34. https://www.theguardian.com/technology/2019/jan/20/shoshana-zuboff-age-of-surveillance-capitalism-google-facebook; http://theconversation.com/explainer-what-is-surveillance-capitalism-and-how-does-it-shape-our-economy-119158

35. Although this exact wording was the title of an article in *The Economist* (https://www.economist.com/leaders/2017/05/06/the-worlds-most-valuable-resource-is-no-longer-oil-but-data), the phrase 'Data is the new oil' is originally attributed to Clive Humby, a British mathematician and data scientist. When Humby first introduced the phrase in 2006, he used it in the context of explaining why data science is important: 'Data is the new oil. It's valuable, but if unrefined it cannot really be used. It has to be changed into gas, plastic, chemicals, etc to create a valuable entity that drives profitable activity; so must data be broken down, analyzed for it to have value.'

36. Brandtzaeg, P. B., A. Pultier, and G. M. Moen. 'Losing control to data-hungry apps: a mixed-methods approach to mobile app privacy.' *Social Science Computer Review* 37(4) (2019): 466–88.

37. https://theconversation.com/7-in-10-smartphone-apps-share-your-data-with-third-party-services-72404

38. Englehardt, S., and A. Narayanan. 'Online tracking: a 1-million-site measurement and analysis.' In *Proceedings of the 2016 ACM SIGSAC Conference on Computer and Communications Security* (2016): 1388–401.

39. https://www.buzzfeednews.com/article/azeenghorayshi/grindr-hiv-status-privacy

40. MacMillan, Douglas. 'App developers gain access to millions of Gmail inboxes – Google and others enable scanning of emails by data miners.' *Wall Street Journal*, Eastern edition, New York, 3 July 2018: A.1.

41. For examples of sources that discuss the symbiotic relationship between the data collecting economy and AI, see: Zillner, Sonja, et al. 'Data

economy 2.0: from big data value to AI value and a European data space.' In *The Elements of Big Data Value: Foundations of the Research and Innovation Ecosystem*, ed. Edward Curry et al. Springer, 2021: 379–99; see also the section on how big data contributes to AI in https://www.sec.gov/news/speech/bauguess-big-data-ai

42. https://online.maryville.edu/blog/big-data-is-too-big-without-ai/

43. https://www.theguardian.com/news/2018/may/06/cambridge-analytica-how-turn-clicks-into-votes-christopher-wylie; https://qz.com/1232873/what-can-politicians-learn-from-tracking-your-psychology-pretty-much-everything/

44. https://www.ft.com/content/d3bd46cb-75d4-40ff-a0cd-6d7f33d58d7f

45. Aitken, R. '"All data is credit data": constituting the unbanked.' *Competition and Change* 21(4) (2017): 274–300; Hurley, M., and J. Adebayo, 'Credit scoring in the era of big data.' *Yale Journal of Law and Technology* 18(1) (2017): 5.

46. https://fortune.com/2022/06/28/after-roe-v-wade-fear-of-a-i-surveillance-abortion/

47. https://www.lexology.com/library/detail.aspx?g=557f9fd6-a8b3-48de-a402-08a21c279c4d

48. https://www.forbes.com/sites/kashmirhill/2013/12/19/data-broker-was-selling-lists-of-rape-alcoholism-and-erectile-dysfunction-sufferers/#207320161d53

49. https://www.forbes.com/sites/kashmirhill/2013/12/19/data-broker-was-selling-lists-of-rape-alcoholism-and-erectile-dysfunction-sufferers/#207320161d53

50. US Senate Committee on Commerce, Science, and Transportation. 'A review of the data broker industry: collection, use, and sale of consumer data for marketing purposes' (18 December 2013). Available at: http://www.commerce.senate.gov/public/?a=Files.Serve&;File_id=0d2b3642-6221-4888-a631-08f2f255b577

51. https://web.archive.org/web/20180731211011/https://business.weather.com/writable/documents/Financial-Markets/InvestorInsights_SolutionSheet.pdf; https://arstechnica.com/tech-policy/2019/01/weather-channel-app-helped-advertisers-track-users-movements-lawsuit-says/

52. Facebook: https://www.nytimes.com/2018/12/18/us/politics/facebook-data-sharing-deals.html; Twitter: https://help.twitter.com/en/safety-and-security/data-through-partnerships; Tiktok: https://www.cnbc.com/2022/02/08/

tiktok-shares-your-data-more-than-any-other-social-media-app-study.html; Paypal: https://www.paypal.com/ie/webapps/mpp/ua/third-parties-list

53. https://www.paypal.com/us/webapps/mpp/ua/privacy-full#dataCollect

54. As estimated from the policies published in 2008 from 75 of the most popular websites, assuming most people read 250 words per minute; McDonald, Aleecia M., and Lorrie Faith Cranor. 'The cost of reading privacy policies.' *I/S: A Journal of Law and Policy for the Information Society* 4(3) (2008): 543–62.

55. Reidenberg, Joel R., et al. 'Disagreeable privacy policies: mismatches between meaning and users' understanding.' *Berkeley Technology Law Journal* 30(1) (2015): 39.

56. Obar, Jonathan A., and Anne Oeldorf-Hirsch. 'The biggest lie on the internet: ignoring the privacy policies and terms of service policies of social networking services.' *Information, Communication and Society* 23(1) (2020): 128–47.

57. This estimate was created in 2008 so is severely outdated, but the basic point still holds. McDonald, Aleecia M., and Lorrie Faith Cranor. 'The cost of reading privacy policies.' *I/S: A Journal of Law and Policy for the Information Society* 4(3) (2008): 543–62.

58. Korunovska, Jana, Bernadette Kamleitner, and Sarah Spiekermann. 'The challenges and impact of privacy policy comprehension.' arXiv preprint arXiv:2005.08967 (2020).

59. https://uxdesign.cc/dark-patterns-in-ux-design-7009a83b233c; https://techcrunch.com/2018/07/01/wtf-is-dark-pattern-design/

60. https://www.notebookcheck.net/New-study-finds-60-of-apps-used-by-U-S-schools-share-student-data-with-third-parties-sometimes-without-the-users-knowledge.537473.0.html; https://www.cnn.com/2022/05/26/tech/remote-learning-apps-data-collection/index.html

61. https://venturebeat.com/2018/07/04/google-doesnt-dispute-claims-that-third-party-developers-may-read-your-gmail-messages/

62. https://seleritysas.com/blog/2021/10/20/the-value-of-biometric-data-analytics-for-modern-businesses/; https://www.reuters.com/legal/legalindustry/looking-future-biometric-data-privacy-laws-2022-04-06/

63. https://news.rub.de/english/press-releases/2022-07-07-it-security-how-daycare-apps-can-spy-parents-and-children

64. https://www.dhs.gov/xlibrary/assets/privacy/privacy_advcom_06-2005_testimony_sweeney.pdf

65.   Ohm, Paul. 'Broken promises of privacy: responding to the surprising failure of anonymization.' *UCLA Law Review* 57(6) (2010): 1701–77.

66.   Rocher, Luc, Julien M. Hendrickx, and Yves-Alexandre De Montjoye. 'Estimating the success of re-identifications in incomplete datasets using generative models.' *Nature Communications* 10(1) (2019): 3069.

67.   Vanessa Teague, University of Melbourne, quoted at https://www. theguardian.com/world/2018/jul/13/anonymous-browsing-data-medical-records-identity-privacy

68.   Google: https://www.nytimes.com/2019/11/11/business/google-ascension-health-data.html; Apple: https://support.apple.com/en-in/HT208647; Microsoft: https://www.microsoft.com/en-us/industry/health/enable-personalized-care?rtc=1; Amazon: https://www.cerner.com/blog/cerner-leads-new-era-of-health-care-innovation; IBM: https://www-03.ibm.com/press/us/en/pressrelease/49132.wss

69.   Perez, B., M. Musolesi, and G. Stringhini, 'You are your metadata: identification and obfuscation of social media users using metadata information.' In *Proceedings of the Twelfth International AAAI Conference on Web and Social Media* (2018): 241–50.

70.   De Montjoye, Yves-Alexandre, et al. 'Unique in the crowd: the privacy bounds of human mobility.' *Scientific Reports* 3 (2013): 1376.

71.   https://www.symantec.com/blogs/threat-intelligence/mobile-privacy-apps

72.   https://www.bellingcat.com/resources/articles/2018/07/08/strava-polar-revealing-homes-soldiers-spies/

73.   https://www.techdirt.com/articles/20190723/08540542637/once-more-with-feeling-anonymized-data-is-not-really-anonymous.shtml

74.   Zhu, T., and P. S. Yu. 'Applying differential privacy mechanism in artificial intelligence.' *In 2019 IEEE 39th International Conference on Distributed Computing Systems (ICDCS)* (2019): 1601–9.

75.   Floridi, L. 'What the near future of artificial intelligence could be.' In *The 2019 Yearbook of the Digital Ethics Lab*. Springer, 2020: 127–42; Rankin, D., et al., 'Reliability of supervised machine learning using synthetic data in health care: model to preserve privacy for data sharing.' *JMIR Medical Informatics* 8(7) (2020): e18910.

76.   Rahman, M. S., et al. 'Towards privacy preserving AI based composition framework in edge networks using fully homomorphic encryption.' *Engineering Applications of Artificial Intelligence* 94 (2020): 103737; 'Meet the new twist on data encryption that promises better privacy and security for AI.' VentureBeat (2020).

77. Barnes, S. 'A privacy paradox: social networking in the United States.' *First Monday* 11(9) (2006). http://firstmonday.org/article/view/1394/1312; Kokolakis, S. 'Privacy attitudes and privacy behaviour: a review of current research on the privacy paradox phenomenon.' *Computers and Security* 64 (2017): 122–34.

78. Turow, J., M. Hennessy, and N. Draper. 'The tradeoff fallacy: how marketers are misrepresenting American consumers and opening them up to exploitation.' SSRN preprint 2820060 (2015); https://www.salesforce.com/blog/2016/11/swap-data-for-personalized-marketing.html; https://marketingland.com/survey-99-percent-of-consumers-will-share-personal-info-for-rewards-also-want-brands-to-ask-permission-130786

79. https://www.pewresearch.org/internet/2019/11/15/americans-and-privacy-concerned-confused-and-feeling-lack-of-control-over-their-personal-information/

80. Draper, N. A. 'From privacy pragmatist to privacy resigned: challenging narratives of rational choice in digital privacy debates.' *Policy and Internet* 9(2) (2017): 232–51; Draper, N. A., and J. Turow. 'The corporate cultivation of digital resignation.' *New Media and Society* 21(8) (2019): 1824–39.

81. Marwick, A., and E. Hargittai. 'Nothing to hide, nothing to lose? Incentives and disincentives to sharing information with institutions online.' *Information, Communication and Society* 22(12) (2019): 1697–713.

CHAPTER 4: CAN AI BE FAIR?

1. https://www.nytimes.com/2020/08/20/world/europe/uk-england-grading-algorithm.html; https://www.theguardian.com/education/2020/aug/13/almost-40-of-english-students-have-a-level-results-downgraded

2. https://www.theguardian.com/education/2021/feb/18/the-student-and-the-algorithm-how-the-exam-results-fiasco-threatened-one-pupils-future

3. For these and other cases, see Gebru, Timnit. 'Race and Gender.' In *The Oxford Handbook of Ethics of AI*, ed. Markus D. Dubber, Frank Pasquale, and Sunit Das. Oxford University Press, 2020: 253–270.

4. https://www.theguardian.com/society/2021/nov/09/ai-skin-cancer-diagnoses-risk-being-less-accurate-for-dark-skin-study

5. https://www.theguardian.com/us-news/2023/feb/08/us-immigration-cbp-one-app-facial-recognition-bias

6. Aristotle, *Nicomachean Ethics*, Book V.

7. Examples include unjust contracts and unfair prices. Is it unfair if AI marketing enables stores to charge higher prices? Or if AI leads pharmaceutical manufacturers to set drug prices too high for many needy patients to afford because it predicts that sick people will be willing to pay those high prices? These questions are both fascinating and important, but we will focus on other issues in the text.

8. https://www.statista.com/statistics/191261/number-of-arrests-for-all-offenses-in-the-us-since-1990/

9. Rachlinski, Jeffrey J., et al. 'Does unconscious racial bias affect trial judges.' *Notre Dame Law Review* 84(3) (2008): 1195–296.

10. https://www.nytimes.com/2017/12/20/upshot/algorithms-bail-criminal-justice-system.html

11. American Law Institute, *Model Penal Code: Sentencing*, proposed final draft, 2017: article 6B.09 comment a, 387–9.

12. Kleinberg, Jon, et al. 'Human decisions and machine predictions.' *Quarterly Journal of Economics* 133(1) (2018): 237–93. All statistics in these sections come from this source unless otherwise indicated.

13. Dror, Itiel E. 'Cognitive and human factors in expert decision making: six fallacies and the eight sources of bias.' *Analytical Chemistry* 92(12) (2020): 7998–8004.

14. Zeng, Jiaming, Berk Ustun, and Cynthia Rudin. 'Interpretable classification models for recidivism prediction.' *Journal of the Royal Statistical Society: Series A (Statistics in Society)* (2017): 689–722.

15. Dressel, Julia, and Hany Farid. 'The accuracy, fairness, and limits of predicting recidivism.' *Science Advances* 4(1) (2018): eaao5580.

16. Angwin, Julia, et al. 'Machine bias.' *ProPublica*, 23 May 2016. https://www.propublica.org/article/machine-bias-risk-assessments-in-criminal-sentencing

17. Dietrich, W., C. Mendoza, and T. Brennan. 'COMPAS risk scales: demonstrating accuracy equity and predictive parity.' Northpointe technical report (2016). https://www.documentcloud.org/documents/2998391-ProPublica-Commentary-Final070616.html

18. Eva, Ben. 'Algorithmic fairness and base rate tracking.' *Philosophy and Public Affairs* 50(2) (2022): 239–66.

19. Narayanan, Arvind. 'Translation tutorial: 21 fairness definitions and their politics.' In: *Proceedings of the 1st Conference on Fairness, Accountability and Transparency* (2018): 3.

20. Kleinberg, J., S. Mullainathan, and M. Raghavan. 'Inherent trade-offs in the fair determination of risk scores.' Available at https://arxiv.org/abs/1609.05807v2(2016). Kleinberg, J. 'Inherent trade-offs in algorithmic fairness.' In *Abstracts of the 2018 ACM International Conference on Measurement and Modeling of Computer Systems* (2018): 40.

21. See Corbett-Davies, Sam, et al. 'Algorithmic decision making and the cost of fairness.' In *Proceedings of the 23rd ACM SIGKDD International Conference on Knowledge Discovery and Data Mining* (2017): 797–806.

22. Miller, Andrea L. 'Expertise fails to attenuate gendered biases in judicial decision-making.' *Social Psychological and Personality Science* 10(2) (2019): 227–34.

23. Harris, Allison P., and Maya Sen. 'Bias and judging.' *Annual Review of Political Science* 22 (2019): 241–59.

24. https://www.propublica.org/article/machine-bias-risk-assessments-in-criminal-sentencing, comments section. Accessed 15 January 2020.

25. Stevenson, Megan. 'Assessing risk assessment in action.' *Minnesota Law Review* 103 (2018): 303–84; https://www.wired.com/story/algorithms-shouldve-made-courts-more-fair-what-went-wrong/

26. Rudin, Cynthia, and Joanna Radin. 'Why are we using black box models in AI when we don't need to? A Lesson from an explainable AI competition.' *Harvard Data Science Review* 1(2) (2019). https://hdsr.mitpress.mit.edu/pub/f9kuryi8/release/8; Rudin, C. 'Stop explaining black box machine learning models for high stakes decisions and use interpretable models instead.' *Nature machine intelligence* 1(5) (2019): 206–15.

27. See 'Attorney General Eric Holder speaks at the National Association of Criminal Defense Lawyers 57th Annual Meeting and 13th State Criminal Justice Network Conference', available at https://www.justice.gov/opa/speech/attorney-general-eric-holder-speaks-national-association-criminal-defense-lawyers-57th

28. Mehrabi, Ninareh, F. et al. 'A survey on bias and fairness in machine learning.' *ACM Computing Surveys* 54(6) (2021): 1–35.

29. Two forms are explained and tested in Yang, Crystal, and Will Dobbie. 'Equal protection under algorithms: a new statistical and legal framework.' *Michigan Law Review* 119(2) (2020): 291–395. A similar method is proposed in Kleinberg, Jon, et al. 'Advances in big data research in economics: algorithmic fairness.' *AEA Papers and Proceedings* (108) (2018) 22–7.

30. Yang, Crystal, and Will Dobbie. 'Equal protection under algorithms: a new statistical and legal framework.' *Michigan Law Review* 119(2) (2020): 291–395.

31. For details on the legal issues in the US, see Yang, Crystal, and Will Dobbie. 'Equal protection under algorithms: a new statistical and legal framework.' *Michigan Law Review* 119(2) (2020): 291–395.

32. See *Loomis v. Wisconsin*, 881 N.W.2d 749 (Wis. 2016), cert. denied, 137 S.Ct. 2290 (2017).

33. Washington, Anne L. 'How to argue with an algorithm: lessons from the COMPAS-ProPublica debate.' *Colorado Technology Law Journal* 17(1) (2018): 131.

34. https://www.wired.com/story/ai-experts-want-to-end-black-box-algorithms-in-government/. We owe many of our points here to illuminating discussions with Kate Vredenburgh.

35. Rudin, C. 'Stop explaining black box machine learning models for high stakes decisions and use interpretable models instead.' *Nature machine intelligence* 1(5) (2019): 206–215.

36. This paragraph is deeply indebted to: Zhou, Yishan, and David Danks. 'Different "intelligibility" for different folks.' IN *AIES '20: Proceedings of the AAAI/ACM Conference on AI, Ethics, and Society* (2020): 194–9.

37. https://arstechnica.com/tech-policy/2019/09/algorithms-should-have-made-courts-more-fair-what-went-wrong/?comments=1& comments-page=1

38. Alufaisan, Yasmeen, et al. 'Does explainable artificial intelligence improve human decision-making?' In *Proceedings of the AAAI Conference on Artificial Intelligence* 35(8) (2021): 6618–26; see also Schemmer, Max, et al. 'A meta-analysis of the utility of explainable artificial intelligence in human-AI decision-making.' In *Proceedings of the 2022 AAAI/ACM Conference on AI, Ethics, and Society* (2022): 617–26.

39. For example, see Saleiro, P., et al. 'Aequitas: A bias and fairness audit toolkit.' arXiv preprint arXiv:1811.05577 (2018).

40. For example, see https://lacunafund.org/health/, or https://gretel.ai/blog/automatically-reducing-ai-bias-with-synthetic-data

41. For example, see https://fairplay.ai/

42. For example, see Madaio, Michael A., et al. 'Co-designing checklists to understand organizational challenges and opportunities around fairness in AI.' In *Proceedings of the 2020 CHI Conference on Human Factors in Computing Systems* (2020): 1–14.

43. For example, Ruf, B., and M. Detyniecki. 'A tool bundle for AI fairness in practice.' In *Extended Abstracts of the 2022 CHI Conference on Human Factors in Computing Systems* (2022): 1–3.

44. For some additional strategies being developed, see Chen, R. J., et al. 'Algorithm fairness in AI for medicine and healthcare.' arXiv preprint arXiv:2110.00603 (2021).

45. Schaich Borg, Jana. 'The AI field needs translational ethical AI research.' *AI Magazine* 43(3) (2022): 294–307.

46. Schaich Borg, Jana. 'Four investment areas for ethical AI: transdisciplinary opportunities to close the publication-to-practice gap.' *Big Data and Society* 8(2) (2021): 20539517211040197.

CHAPTER 5: CAN AI (OR ITS CREATORS OR USERS) BE RESPONSIBLE?

1. National Transportation Safety Board. 'Highway accident report: collision between vehicle controlled by developmental automated driving system and pedestrian, Tempe, Arizona, March 18, 2018.' NTSB/HAR-19/03 PB2019-101402 (2019). Videos of Vasquez driving and of Herzberg being hit can also be seen on YouTube, though they might be disturbing.

2. https://www.wired.com/story/uber-self-driving-car-fatal-crash/

3. https://www.phoenixnewtimes.com/news/uber-self-driving-crash-arizona-vasquez-wrongfully-charged-motion-11583771

4. https://cleantechnica.com/2022/06/13/ubers-deadly-2018-autonomous-vehicle-crash-isnt-over-yet/; https://www.phoenixnewtimes.com/news/uber-self-driving-crash-arizona-vasquez-wrongfully-charged-motion-11583771

5. https://www.wired.com/story/uber-self-driving-car-fatal-crash/

6. https://www.wired.com/story/uber-self-driving-car-fatal-crash/

7. https://www.theinformation.com/articles/the-uber-whistleblowers-email

8. National Transportation Safety Board. 'Highway accident report: collision between vehicle controlled by developmental automated driving system and pedestrian, Tempe, Arizona, March 18, 2018.' NTSB/HAR-19/03 PB2019-101402 (2019).

9. National Transportation Safety Board. 'Highway accident report: collision between vehicle controlled by developmental automated driving system and pedestrian, Tempe, Arizona, March 18, 2018.' NTSB/HAR-19/03 PB2019-101402 (2019).

10. https://techcrunch.com/2018/03/29/uber-has-settled-with-the-family-of-the-homeless-victim-killed-last-week/

11. https://www.phoenixnewtimes.com/news/uber-self-driving-crash-arizona-vasquez-wrongfully-charged-motion-11583771

12. National Transportation Safety Board. 'Highway accident report: collision between vehicle controlled by developmental automated driving system and pedestrian, Tempe, Arizona, March 18, 2018.' NTSB/HAR-19/03 PB2019-101402 (2019).

13. https://cleantechnica.com/2022/06/13/ubers-deadly-2018-autonomous-vehicle-crash-isnt-over-yet/; https://www.phoenixnewtimes.com/news/uber-self-driving-crash-arizona-vasquez-wrongfully-charged-motion-11583771

14. https://www.wired.com/story/uber-self-driving-car-fatal-crash/

15. https://www.ntsb.gov/news/press-releases/Pages/NR20191119c.aspx

16. https://www.npr.org/sections/thetwo-way/2018/03/27/597331608/arizona-suspends-ubers-self-driving-vehicle-testing-after-fatal-crash#:~:text=%22We%20decided%20to%20not%20reapply,in%20cities%20around%20the%20country

17. https://sfist.com/2016/12/15/uber_blames_humans_as_more_reports/

18. https://www.theverge.com/2016/12/22/14062926/uber-self-driving-car-move-arizona-san-francisco-dmv

19. https://www.theguardian.com/technology/2016/dec/21/uber-cancels-self-driving-car-trial-san-francisco-california

20. https://www.theguardian.com/technology/2018/mar/28/uber-arizona-secret-self-driving-program-governor-doug-ducey

21. https://www.phoenixnewtimes.com/news/arizona-governor-doug-ducey-shares-blame-fatal-uber-crash-10319379; https://www.theguardian.com/technology/2018/mar/28/uber-arizona-secret-self-driving-program-governor-doug-ducey

22. https://www.phoenixnewtimes.com/news/arizona-governor-doug-ducey-shares-blame-fatal-uber-crash-10319379

23. https://www.azcentral.com/story/news/local/tempe/2019/03/19/arizona-city-tempe-sued-family-uber-self-driving-car-crash-victim-elaine-herzberg/3207598002/

24. Note that condemning and ostracizing an AI in these ways we are describing is more than just expressing sadness about the harm the AI caused. The condemnation and ostracization would be applied because we ascribe fault to the AI and punish it because we do not think it is safe enough.

25. https://www.wsj.com/articles/google-ai-chatbot-bard-chatgpt-rival-bing-a4c2d2ad; https://www.washingtonpost.com/technology/2023/01/27/chatgpt-google-meta/

26. https://www.govtech.com/policy/the-battle-over-california-social-media-liability-bill-mounts; https://www.politico.com/newsletters/digital-future-daily/2022/06/29/small-fry-ai-dc-try-00043278

CHAPTER 6: CAN AI INCORPORATE HUMAN MORALITY?

1. Bostrom, N. *Superintelligence: Paths, Dangers, Strategies.* Oxford University Press, 2014.

2. Awad, E., et al. 'Computational ethics.' *Trends in Cognitive Sciences* 26(5) (2022): 388–405; Russell, Stuart. *Human Compatible: Artificial Intelligence and the Problem of Control.* Penguin 2019.

3. Asimov, I. 'Runaround' (1950). Reprinted in *I, Robot.* Doubleday.

4. Kant, I. 'On a supposed right to lie from altruistic motives' (1797). In *Immanuel Kant: Critique of Practical Reason and Other Writings in Moral Philosophy,* trans. Lewis White Beck. University of Chicago Press, 1949; reprint: Garland Publishing Company, 1976.

5. These phenomena are well documented in the moral judgment field. As a few examples: McDonald, Kelsey, et al. 'Valence framing effects on moral judgments.' *Cognition* 212 (2021): 104703; Rehren, Paul, and Walter Sinnott-Armstrong. 'Moral framing effects within subjects.' *Philosophical Psychology* 34(5) (2021): 611–36; Rehren, Paul, and Walter Sinnott-Armstrong. 'How Stable are moral judgments?' *Review of Philosophy and Psychology* (2022).

6. https://www.organdonor.gov/learn/organ-donation-statistics

7. Strohmaier, Susanne, et al. 'Survival benefit of first single-organ deceased donor kidney transplantation compared with long-term dialysis across ages in transplant-eligible patients with kidney failure.' *JAMA Network Open* 5(10) (2022): e2234971.

8. https://www.kidney.org/news/newsroom/factsheets/Organ-Donation-and-Transplantation-Stats

9. https://qz.com/1383083/how-ai-changed-organ-donation-in-the-us

10. Seyahi, Nurhan, and Seyda Gul Ozcan. 'Artificial intelligence and kidney transplantation.' *World Journal of Transplantation* 11(7) (2021): 277–89; Schwantes, Issac R., and David A. Axelrod. 'Technology-enabled care and

artificial intelligence in kidney transplantation.' *Current Transplantation Reports* 8(3) (2021): 235–40.

11. Cf. Chan, Lok, et al. 'Which features of patients are morally relevant in ventilator triage? A survey of the UK public.' *BMC Medical Ethics* 23 (2022): 33; Freedman, R., et al. 'Adapting a kidney exchange algorithm to align with human values.' *Artificial Intelligence* 283 (2020), 103261 (pp. 1-14).

12. Cf. Chan, Lok, et al. 'Which features of patients are morally relevant in ventilator triage? A survey of the UK public.' *BMC Medical Ethics* 23 (2022): 33.

13. McElfresh, Duncan C., et al. 'Indecision modeling.' In *Proceedings of the AAAI Conference on Artificial Intelligence* 35(7) (2021): 5975–83.

14. Conitzer, V., 'Designing preferences, beliefs, and identities for artificial intelligence.' In *Proceedings of the AAAI Conference on Artificial Intelligence* 33(1) (2019): 9755–59; Noothigattu, R., et al. 'A voting-based system for ethical decision making.' In *Proceedings of the AAAI Conference on Artificial Intelligence* 32(1) (2018): 1587–94; Zhang, H. and V. Conitzer. 'A PAC framework for aggregating agents' judgments.' In *Proceedings of the AAAI Conference on Artificial Intelligence* 33(1) (2019): 2237–44; Elkind, Edith, and Arkadii Slinko. 'Rationalizations of voting rules.' *Handbook of Computational Social Choice*, ed. Felix Brandt. Cambridge University Press, 2016: 169–96.

15. T. Marshall, dissenting opinion, *Gregg v. Georgia*, 428 U.S. 153 (1976), 428, referring to *Furman v. Georgia* 408 U.S. 238 (1972) 360–69.

16. Sarat, A., and N. Vidmar. 'Public opinions, the death penalty, and the Eighth Amendment: testing the Marshall hypothesis.' *Wisconsin Law Review* (1976): 171–206.

17. Provenzano, Michele, et al. 'Smoking habit as a risk amplifier in chronic kidney disease patients.' *Scientific Reports* 11(1) (2021): 14778; Van Laecke, Steven, and Wim Van Biesen. 'Smoking and chronic kidney disease: seeing the signs through the smoke?' *Nephrology Dialysis Transplantation* 32(3) (2017): 403–5; Fan, Zhenliang, et al. 'Alcohol consumption can be a "double-edged sword" for chronic kidney disease patients.' *Medical Science Monitor* 25 (2019): 7059–72.

18. Rawls, John. *A Theory of Justice*. Harvard University Press, 1971.

19. Sinnott-Armstrong, Walter. 'Idealized observer theories in ethics.' In *Oxford Handbook of Ethical Theory*, second edition, ed. David Copp and Connie Rosati. Oxford University Press, forthcoming.

CHAPTER 7: WHAT CAN WE DO?

1. See, for example, the annual Association for Computing Machinery Conference on Fairness, Accountability, and Transparency (ACM FAccT) or the annual Association for the Advancement of Artificial Intelligence Conference on AI, Ethics, and Society (AAAI/ACM AIES).

2. Winfield, A. 'An updated round up of ethical principles of robotics and AI', 18 April 2019. https://alanwinfield.blogspot.com/2019/04/an-updated-round-up-of-ethical.html; Hagendorff, T. 'The ethics of AI ethics: an evaluation of guidelines.' *Minds and Machines* 30(1) (2020): 99–120; Ryan, M., and B. C. Stahl. 'Artificial intelligence ethics guidelines for developers and users: clarifying their content and normative implications.' *Journal of Information, Communication and Ethics in Society* 19(1) (2021): 61–86; Morley, J., et al. 'From what to how: an initial review of publicly available AI ethics tools, methods and research to translate principles into practices.' *Science and Engineering Ethics* 26(4) (2021): 2141–68; Jobin, A., M. Ienca, and E. Vayena. 'The global landscape of AI ethics guidelines.' *Nature Machine Intelligence* 1(9) (2019): 389–99.

3. Morley, J., et al., 'From what to how: an initial review of publicly available AI ethics tools, methods and research to translate principles into practices.' *Science and Engineering Ethics* 26(4) (2021): 2141–68; Schiff, D., et al. 'Principles to practices for responsible AI: closing the gap.' arXiv preprint arXiv:2006.04707 (2020); McNamara, A., J. Smith, and E. Murphy-Hill. 'Does ACM's code of ethics change ethical decision making in software development?' In *Proceedings of the 2018 26th ACM Joint Meeting on European Software Engineering Conference and Symposium on the Foundations of Software Engineering* (2018): 729–33.

4. https://www.pwc.com/gx/en/issues/data-and-analytics/artificial-intelligence/what-is-responsible-ai.html

5. Bellamy, R. K., et al. 'AI Fairness 360: an extensible toolkit for detecting and mitigating algorithmic bias.' *IBM Journal of Research and Development* 63(4/5) (2019): 4:1–4:15.

6. Bird, S., et al. 'Fairlearn: a toolkit for assessing and improving fairness in AI.' Microsoft technical report MSR-TR-2020-32 (2020).

7. Seng, M., A. Lee and J. Singh. 'The landscape and gaps in open source fairness toolkits.' In *Proceedings of the 2021 CHI Conference on Human Factors in Computing Systems* (2021): 1–13.

8. Saha, D., et al. 'Measuring non-expert comprehension of machine learning fairness metrics.' *Proceedings of the 37th International Conference on*

Machine Learning, in Proceedings of Machine Learning Research 119 (2020): 8377–87.

9. Dimensional Research. 'Artificial intelligence and machine learning projects are obstructed by data issues: global survey of data scientists, AI experts and stakeholders' (May 2019).

10. Schmelzer, R. 'The changing venture capital investment climate for AI.' *Forbes*, 9 August 2020.

11. https://www.statista.com/statistics/943151/ai-funding-worldwide-by-quarter/#:~:text=In%20the%20third%20quarter%20of,to%20208%20billion%20U.S.%20dollars.

12. Womack, J. P., and D. T. Jones. 'Lean thinking: banish waste and create wealth in your corporation.' *Journal of the Operational Research Society* 48(11) (1997): 1148; https://www.agilealliance.org/agile101/12-principles-behind-the-agile-manifesto/; Abbas, Noura, Andrew M. Gravell, and Gary B. Wills. 'Historical roots of agile methods: where did "agile thinking" come from?' In *International Conference on Agile Processes and Extreme Programming in Software Engineering* (2008): 94–103.

13. Lean philosophies and methodologies are not the same as agile philosophies and methodologies, but they have a lot of overlap. Thus, for our purposes, we refer to them collectively as 'lean-agile'. Petersen, Kai. 'Is lean agile and agile lean? A comparison between two software development paradigms.' In *Modern Software Engineering Concepts and Practices: Advanced Approaches,* ed. A. H. Dogru and V. Biçer. IGI Global, 2011: 19–46.

14. Mersino, A. 'Agile project success rates are 2x higher than traditional projects.' *Medium*, 25 May 2020.

15. https://digital.ai/resource-center/analyst-reports/state-of-agile-report

16. https://www.rackspace.com/sites/default/files/pdf-uploads/Rackspace-White-Paper-AI-Machine-Learning.pdf

17. Beck, Ulrich. *Gegengifte: Die organisierte Unverantwortlichkeit.* Suhrkamp, 1988. This quotation was brought to our attention by Hagendorff, T. 'The ethics of AI ethics: an evaluation of guidelines.' *Minds and Machines* 30(1) (2020): 99–120.

18. Rakova, B., et al. 'Where responsible AI meets reality: practitioner perspectives on enablers for shifting organizational practices.' *Proceedings of the ACM on Human–Computer Interaction* 5(CSCW1) (2021): 7; Findlay, M., and J. Seah. 'An ecosystem approach to ethical AI and data use: experimental reflections.' In *2020 IEEE/ITU International Conference on Artificial Intelligence for Good (AI4G)* (2020): 192–7. IEEE.

19. Rakova, B., et al. 'Where responsible AI meets reality: practitioner perspectives on enablers for shifting organizational practices.' *Proceedings of the ACM on Human–Computer Interaction* 5(CSCW1) (2021): 7.

20. Rakova, B., et al. 'Where responsible AI meets reality: practitioner perspectives on enablers for shifting organizational practices.' *Proceedings of the ACM on Human–Computer Interaction* 5(CSCW1) (2021): 7; Findlay, M., and J. Seah. 'An ecosystem approach to ethical AI and data use: experimental reflections.' In *2020 IEEE/ITU International Conference on Artificial Intelligence for Good (AI4G)* (2020): 192–7.

21. Microsoft: https://www.theverge.com/2023/3/13/23638823/microsoft-ethics-society-team-responsible-ai-layoffs; Twitter: https://www.wired.com/story/twitter-ethical-ai-team/; Meta: https://fortune.com/2022/09/09/meta-axes-responsible-innovation-team-downside-to-products/; Google: https://www.ft.com/content/26372287-6fb3-457b-9e9c-f722027f36b3

22. Vakkuri, V., et al. '"This is just a prototype": how ethics are ignored in software startup-like environments.' In Stray, V., R. Hoda, M. Paasivaara, and P. Kruchten (eds), *Agile Processes in Software Engineering and Extreme Programming: 21st International Conference on Agile Software Development, XP 2020.* (2020): 195–210; https://www.forbes.com/sites/lanceeliot/2022/04/21/ai-startups-finally-getting-onboard-with-ai-ethics-and-loving-it-including-those-newbie-autonomous-self-driving-car-tech-firms-too/?sh=418911adb9c6

23. El-Zein, A. 'As engineers, we must consider the ethical implications of our work', *Guardian*, 5 December 2013.

24. Arledge, C. 'Design ethics and the limits of the ethical designer', 2 January 2019. https://www.viget.com/articles/design-ethics-and-the-limits-of-the-ethical-designer/

25. https://everything2.com/title/The+Official+%2522Not-it%2522+Rules

26. A poll of over 100,000 software developers on StackOverflow, an online community for software developers, illustrates this lack of consensus well. When asked, 'Do developers have an obligation to consider the ethical implications of their code?', 80 per cent said yes, 14 per cent said they were unsure, and 6 per cent said no. When asked, 'Who is ultimately most responsible for code that accomplishes something unethical?' 57 per cent chose upper management, 23 per cent chose the person who came up with the idea, and 20 per cent chose the developer who wrote it. When asked, 'Who is primarily responsible for considering the ramifications of AI?' 48 per cent chose 'the developers or the people creating the AI',

28 per cent chose 'a governmental or other regulatory body', 17 per cent chose 'prominent industry leaders', and almost 8 per cent chose 'nobody'. https://insights.stackoverflow.com/survey/2018#overview

27. Schwab, K. 'Designers: you're policy makers. It's time to act like it.' *Fast Company*, 28 February 2018.

28. Note, however, that efforts like ethical checklists do try to provide some guidance about how to implement moral AI principles in AI development teams' practices. They just typically don't address many of the cultural issues that may prevent those checklists from being used most effectively.

29. For a much deeper discussion of this topic, see Schaich Borg, Jana, 'The AI field needs translational ethical AI research.' *AI Magazine* 43(3) (2022): 294–307.

30. Examples include: https://resources.sei.cmu.edu/asset_files/ FactSheet/2019_010_001_636622.pdf; https://www.microsoft.com/en-us/ research/project/ai-fairness-checklist/; and https://deon.drivendata.org/

31. https://ai.googleblog.com/2022/11/the-data-cards-playbook-toolkit-for.html

32. e.g. https://techmonitor.ai/technology/ai-and-automation/ ai-auditing-next-big-thing-will-it-ensure-ethical-algorithms

33. Harrison, J. S., R. A. Phillips, and R. E. Freeman. 'On the 2019 Business Roundtable "Statement on the purpose of a corporation".' *Journal of Management* 46(7) (2020): 1223–37.

34. Buhmann, A., and C. Fieseler. 'Towards a deliberative framework for responsible innovation in artificial intelligence.' *Technology in Society* 64 (2021): 101475. For example, see: Seng, M., A. Lee, and J. Singh. 'The landscape and gaps in open source fairness toolkits.' In *Proceedings of the 2021 CHI Conference on Human Factors in Computing Systems* (2021): 1–13.

35. Rakova, B., et al. 'Where responsible AI meets reality: practitioner perspectives on enablers for shifting organizational practices.' *Proceedings of the ACM on Human–Computer Interaction* 5(CSCW1) (2021): 7.

36. Thompson, D. F. 'What is practical ethics?' In *Ethics at Harvard, 1987–2007*, Edmond J. Safra Center for Ethics, Harvard University, 2007.

37. De Cremer, D., and C. Moore. 'Toward a better understanding of behavioral ethics in the workplace.' *Annual Review of Organizational Psychology and Organizational Behavior* 7(1) (2020): 369–93.

38. Smith, I. H., and M. Kouchaki. 'Ethical learning: the workplace as a moral laboratory for character development.' *Social Issues and Policy Review* 15(1) (2021): 277–322.

39. Findlay, M., and J. Seah. 'An ecosystem approach to ethical AI and data use: experimental reflections.' In 2020 *IEEE/ITU International Conference on Artificial Intelligence for Good (AI4G)* (2020): 192–7; Smith, I. H., and M. Kouchaki. 'Ethical learning: the workplace as a moral laboratory for character development.' *Social Issues and Policy Review* 15(1) (2021): 277–322.

40. Chugh, D., and M. C. Kern. 'Ethical learning: releasing the moral unicorn.' In *Organizational Wrongdoing: Key Perspectives and New Directions*, ed. D. Palmer, K. Smith- Crowe, and R. Greenwood. Cambridge University Press, 2016: 474–503.

41. Metcalf, J., and E. Moss. 'Owning ethics: corporate logics, Silicon Valley, and the institutionalization of ethics.' *Social Research* 86(2) (2019): 449–76.

42. McLennan, S., et al. 'An embedded ethics approach for AI development.' *Nature Machine Intelligence* 2(9) (2020): 488–90.

43. Synced. 'Exploring gender imbalance in AI: numbers, trends, and discussions', 13 March 2020. https://syncedreview.com/2020/03/13/exploring-gender-imbalance-in-ai-numbers-trends-and-discussions/#:~:text=%E2%80%93%2012%25,machine%20learning%20conferences%20in%202017; D'Ambra, L. 'Women in artificial intelligence: a visual study of leadership across industries.' Emerj Artificial Intelligence Research (2017). https://emerj.com/ai-market-research/women-in-artificial-intelligence-visual-study-leaderships-across-industries/; World Economic Forum, 'Assessing Gender Gaps in Artificial Intelligence.' In *The Global Gender Gap Report* (2018): 29–31.

44. Smith, B., and C. A. Browne. 'Tech firms need more regulation.' *The Atlantic*, 9 September 2019.

45. The roles of UX professionals are constantly in flux, but when UX researcher and designer roles are separated, UX researchers are typically tasked with capturing user needs, pains, and behaviours through different observation and feedback collection methods, while UX designers are typically tasked with crafting effective interfaces and experiences based on what UX researchers learn. Especially in smaller companies, though, these responsibilities may be assigned to a single UX role called 'UX designer'.

46. Chou, T. 'A leading Silicon Valley engineer explains why every tech worker needs a humanities education.' *Quartz*, 28 June 2017.

47. Sterman, J. D., 'Learning from evidence in a complex world.' *American Journal of Public Health* 96(3) (2006): 505–14.

48. Boyatzis, R., K. Rochford, and K. V. Cavanagh. 'Emotional intelligence competencies in engineer's effectiveness and engagement.' *Career Development International* 22(1) (2017): 70–86.

49. Tanner, C., and M. Christen. 'Moral intelligence: framework for understanding moral competences.' In *Empirically Informed Ethics: Morality between Facts and Norms,* ed. M. Christen et al. Springer, 2014: 119–136.

50. Martin, Diana Adela, Eddie Conlon, and Brian Bowe. 'A multi-level review of engineering ethics education: towards a socio-technical orientation of engineering education for ethics.' *Science and Engineering Ethics* 27(5) (2021): 1–38.

51. Antes, A. L., et al. 'Evaluating the effects that existing instruction on responsible conduct of research has on ethical decision making.' *Academic Medicine* 85(3) (2010): 519–26.

52. Cech, E. A. 'Culture of disengagement in engineering education?' *Science, Technology, and Human Values* 39(1) (2014): 42–72.

53. Harris, C. E., Jr, et al. *Engineering Ethics: Concepts and Cases.* Cengage Learning, 2013.

54. Examples: Hertz, S. G., and T. Krettenauer. 'Does moral identity effectively predict moral behavior? A meta-analysis.' *Review of General Psychology* 20(2) (2016): 129–40; Gu, J., and C. Neesham. 'Moral identity as leverage point in teaching business ethics.' *Journal of Business Ethics* 124(3) (2014): 527–36; Miller, G. 'Aiming professional ethics courses toward identity development.' In *Ethics Across the Curriculum: Pedagogical Perspectives*, ed. E.E. Englehardt and M.S. Pritchard. Springer, 2018: 89–105; Pierrakos, O., et al. 'Reimagining engineering ethics: from ethics education to character education.' In *2019 IEEE Frontiers in Education Conference (FIE)* (2019): 1–9; Han, H. 'Virtue ethics, positive psychology, and a new model of science and engineering ethics education.' *Science and Engineering Ethics* 21(2) (2015): 441–60; Stelios, S. 'Professional engineers: interconnecting personal virtues with human good.' *Business and Professional Ethics Journal* 39(2) (2020): 253–68.

55. Saltz, J., et al. 'Integrating ethics within machine learning courses.' *ACM Transactions on Computing Education* 19(4) (2019): 32; Schwab, K. 'Mozilla's ambitious plan to teach coders not to be evil.' *Fast Company*, 15 October 2018.

56. Think this sounds impossible? Seventy-two-year-old US Congressman Don Beyer has been getting a Master's degree in machine learning while performing his congressional duties to prepare him to craft AI

legislation more effectively, so motivated chief learning officers could do something similar. See https://www.techdirt.com/2023/01/05/congressman-moonlighting-as-a-masters-degree-student-in-ai/

57. Wagner, K. 'Mark Zuckerberg says he's "fundamentally uncomfortable" making content decisions for Facebook.' Vox, 2018. https://www.vox.com/2018/2013/2022/17150772/mark-zuckerberg-facebook-content-policy-guidelines-hate-free-speech.

58. https://www.speakupaustin.org/

59. https://www.dublincity.ie/business/economic-development-and-enterprise/economic-development/your-dublin-your-voice

60. https://www.citizenlab.co/about

61. Belone, L.,et al. 'Community-based participatory research conceptual model: community partner consultation and face validity.' *Qualitative Health Research* 26(1) (2016): 117–35.

62. https://int.nyt.com/data/documenthelper/639-clarifai-letter/3cd943d873d78c7cdcdc/optimized/full.pdf#page=1; Metz, C. 'Is ethical A.I. even possible?' *New York Times*, 1 March 2019.

63. https://www.uschamber.com/technology/in-the-global-race-to-lead-on-artificial-intelligence-america-must-win

64. For example, the number of patent filings rose significantly after the European Union and US approved emission or biofuel regulations, and patent filings actually decreased when regulations were loosened (Johnstone, N. et al. 'Environmental policy stringency and technological innovation: evidence from survey data and patent counts.' *Applied Economics* 44(17) (2012): 2157–70). Within the data privacy sector, start-up accelerator managers, lawyers specializing in data protection/information technology law and entrepreneurship, and private venture capital partners report that Europe's General Data Protection Regulations increased the amount of innovation and number of new start-ups they encountered (Martin, N. et al. 'How data protection regulation affects startup innovation.' *Information Systems Frontiers* 21(6) (2019): 1307–24). As a different angle on the issue, teams are more creative and innovative when they have moderate levels of design constraints on what they produce (Acar, O. A., M. Tarakci, and D. Van Knippenberg. 'Creativity and innovation under constraints: a cross-disciplinary integrative review.' *Journal of Management* 45(1) (2019): 96–121).

65. Kharpal, A. 'Commentary: Big Tech's calls for more regulation offers a chance for them to increase their power.' Business & Human Rights Resource Centre, 26 January 2020.

66. Bennear, L. S., and J. B. Wiener. 'Adaptive regulation: instrument choice for policy learning over time.' Draft working paper, 2019.

67. Clark, J., and G. K. Hadfield. 'Regulatory markets for AI safety.' arXiv preprint arXiv:2001.00078 (2019).

68. Stix, C., and M. M. Maas. 'Bridging the gap: the case for an "incompletely theorized agreement" on AI policy.' *AI and Ethics* 1(3) (2021): 261–71.

CONCLUSION: IT'S UP TO US

1. United States Atomic Energy Commission. *In the Matter of J. Robert Oppenheimer: Transcript of Hearing before Personnel Security Board, Washington, D. C., April 12, 1954, through, May 6, 1954.* United States Government Printing Office, 1954: 81.

2. https://www.lrb.co.uk/the-paper/v22/n17/steven-shapin/don-t-let-that-crybaby-in-here-again

Index